Emotion and Information Processing

Sachi Nandan Mohanty
Editor

Emotion and Information Processing

A Practical approach

Springer

Editor
Sachi Nandan Mohanty
Department of Computer Science & Engineering
ICFAI Foundation for Higher Education
Hyderabad, India

ISBN 978-3-030-48848-2 ISBN 978-3-030-48849-9 (eBook)
https://doi.org/10.1007/978-3-030-48849-9

This Springer imprint is published by the registered company Springer Nature Switzerland AG
The registered company address is: Gewerbestrasse 11, 6330 Cham, Switzerland

To my mentors, who constantly guided me during my work related to emotions and information processing.
To my loving parents and family members for making me believe that everything is possible if one strongly desires for it.

Preface

It was always my dream to edit a book on emotion, cognition and information processing during my Ph.D. work at IIT Kharagpur, India. Day-to-day human life and activities are governed by two major factors: feelings and thoughts, which we more formally called as emotions and cognitions, respectively. Both the governing factors are considered to be the interior happenings and cannot be monitored by anyone else. The information is being processed in two ways either locally or globally by the human being during the time of decision-making or solving a particular problem. Both emotion and cognition are the most important functions being carried out in the frontal lobes of the brain, hence predominantly assigned to problem-solving and decision-making process.

This book consists of 13 chapters covering many facts like psycho-social intervention on emotional disorders in individuals, impact of emotion and cognition on blended theory, theory and implication of information processing, effects of emotional self-esteem in women, emotional dimension of women in workplace, effects of mental thinking in different age groups irrespective of the gender, negative emotions and its effect on information processing, role of emotions in education and lastly emotional analysis in multi perspective domain adopting machine learning approach.

An additional new development in neuroscience is cognitive neuroscience, which combines the dual approach of understanding the human behaviour is symbolic and physiological according to Simon (1992). Both the approaches are now accepted as normal and natural ways of understanding the underlying cognitive processes in the field of education and administrative behavioural approaches. Emotion and information processing both have expanded an interest in cognition as well as neurological functions in the field of education for facilitating learning and in the field of management for smooth decision-making.

The impression of all the chapters in this book covers diversified domain of cognitive psychology, healthcare sectors, field of education and management sector. The principal audience for this edited book is students and researchers who want to consider emotion and cognition as a topic for future studies and research. As both the topics are widely spread and evolving areas of research nowadays, thus a diversified audience whose curiosity is in emotion and its practical corollaries is embattled.

Bhubaneswar, India Sachi Nandan Mohanty

Reference

Simon, H. A. (1992). What is an explanation of behavior? *Psychological Science, 3*(3), 1–13.

Acknowledgements

I owe my heartfelt thanks to all the people who have enabled me to edit the book and all the authors for contributing their chapters. The completion of this undertaking could not have been possible without the participation of the expert reviewers. So, I would like to offer my thanks to the subject matter experts who could find their time to review the chapters and deliver those in time.

My special thanks to Dr. Suneeta Satpathy, Associate Professor, Department of Computer Science & Engineering, College of Engineering Bhubaneswar, Odisha, India, for her help with proofreading.

I would also like to express my special thanks and gratitude to Ms. Hemalatha Velasasu and Mr. Lilith Dorko for their dedicated support and help in publishing the edited book.

Finally, a ton of thanks to Prof. J. P. Das, University of Alberta, and Prof. Damodar Suar, Department of Humanities and Social Science, Indian Institute of Technology Kharagpur, West Bengal, India, for their constant motivation, encouragement and valuable suggestions as and when needed.

Any attempt at any level cannot be satisfactorily completed without the support and guidance of my parents and friends. So, I would like to thank my parents and friends for constantly encouraging me to make this book possible.

At the end, I am thankful to **Lord Krishna** who blesses me with sound health and abilities and gives courage to perform and complete my work in a successful manner.

Contents

About the Author

Sachi Nandan Mohanty received his Ph.D. from Indian Institute of Technology Kharagpur in the year 2015, with MHRD scholarship from the Government of India. He has recently joined as Associate Professor in the Department of Computer Science and Engineering at ICFAI Foundation for Higher Education, Hyderabad.

Prof. Mohanty's research areas include data mining, big data analysis, cognitive science, fuzzy decision-making, brain–computer interface, and computational intelligence. Prof. S. N. Mohanty has received three Best Paper Awards during his Ph.D. at IIT Kharagpur from International Conference at Beijing, China, and the other at International Conference on Soft Computing Applications organized by IIT Roorkee in the year 2013. He has published 20 SCI Journals. As a Fellow on Indian Society Technical Education (ISTE), The Institute of Engineering and Technology (IET), Computer Society of India (CSI), Member of Institute of Engineers and Senior Member IEEE Computer Society, he is actively involved in the activities of the professional bodies/societies.

He has been bestowed with several awards which include 'Best Researcher Award from Biju Pattnaik University of Technology in 2019', 'Best Thesis Award (first Prize) from Computer Society of India in 2015' and 'Outstanding Faculty in Engineering Award' from the Department of Higher Education, Government of Odisha in 2020. He has received International Travel grant from SERB, Department of Science and Technology, Government of India for chair the session in international conferences in USA in 2020. Ten edited book and Two authored in his credit.

Dr. Mohanty is currently the reviewer of many journals like *Journal of Robotics and Autonomous Systems* (Elsevier), *Computational and Structural Biotechnology* (Elsevier), *Artificial Intelligence Review* (Springer) and *Spatial Information Research* (Springer).

Six edited books published by Wiley, CRC and Springer Nature, and three authored books are on his credit.

Contributors

Rakhi Akhare Computer Science and Engineering Department, Lokmanya Tilak College of Engineering, Navi Mumbai, India

Ahin Banerjee Indian Institute of Technology (BHU), Varanasi, Uttar pradesh, India

Debanshee Datta Indian Institute of Technology (BHU), Varanasi, Uttar pradesh, India

Sanjivani Deokar Computer Science and Engineering Department, Lokmanya Tilak College of Engineering, Navi Mumbai, India

Swati Dhir IMI New Delhi, New Delhi, India

Kanagala Dwarakamai Faculty of Science and Technology, ICFAI Foundation for Higher Education, Hyderabad, India

Richa Goel Amity International Business School, Amity University, Noida, Uttar Pradesh, India

Muskan Gumber CHRIST (Deemed to be University), Bengaluru, India

Anjali Gupta National Institute of Technology, Bhopal, Madhya Pradesh, India

Sanjay K. Gupta Indian Institute of Technology (BHU), Varanasi, India

Rajeev Kumar Indian Institute of Technology Kharagpur, Kharagpur, West bengal, India

Monika Mangla Computer Science and Engineering Department, Lokmanya Tilak College of Engineering, Navi Mumbai, India

Zahida Mariyum Behavioural Coach, ESL Trainer, Bangalore, Karnataka, India

Anjanna Matta Faculty of Science and Technology, ICFAI Foundation for Higher Education, Hyderabad, India

Vaishali Mehta Karnal Institute of Technology and Management, Karnal, India

Preethi Nanjundan Department of Computer Science, Christ (Deemed to be University), Pune, India

Soumya Ranjan Nayak Amity School of Engineering and Technology, Amity University, Noida, Uttar Pradesh, India

Hemanta Kumar Palo Department of Electronics and Communication Engineering, ITER, Siksha 'O' Anusandhan (Deemed to be University), Bhubaneswar, Odisha, India

Smriti Pathak Department of Humanities and Social Sciences, Indian Institute of Technology Kharagpur, Kharagpur, India

Auditi Pramanik Jawaharlal Nehru University, New Delhi, India

Mahima Punni Amity International Business School, Amity University Uttar Pradesh, India

E. Rajesh Kumar Department of Computer Science and Engineering, Koneru Lakshmaiah Education Foundation, Vaddeswaram, AP, India

A. K. V. S. N. Rama Rao Department of Computer Science and Engineering, Koneru Lakshmaiah Education Foundation, Vaddeswaram, AP, India

Seema Sahai Amity International Business School, Amity University Uttar Pradesh, NOIDA, India

Vijay Bhaskar Semwal National Institute of Technology, Bhopal, Madhya Pradesh, India

Anjali Sharma CHRIST (Deemed to be University), Bengaluru, Karnataka, India

Gurinder Singh Amity International Business School, Amity University Uttar Pradesh, Noida, India

Sanjay Kumar Singh Rajendra Institute of Medical Sciences, Ranchi, Jharkhand, India

K. B. L. Srivastava Department of HSS, Indian Institute of Technology Kharagpur, Kharagpur, India

Damodar Suar Indian institute of Technology, Kharagpur, West Bengal, India

G. Sucharitha Faculty of Science and Technology, ICFAI Foundation for Higher Education, Hyderabad, India

Bodepu Tannmayee Faculty of Science and Technology, ICFAI Foundation for Higher Education, Hyderabad, India

Salome Divya Vijaykumar SRM University, Amaravati, AP, India

Chapter 1
Effect of Emotion in Information Processing for Decision-Making

Smriti Pathak and Kailash B. L. Srivastava

Contents

Decision-making is a cognitive process where an individual finally selects a particular course of action, based on his or her values and preferences among several alternative possibilities. It can also be defined as a problem-solving activity, a conscious selection of particular thought or action in a given circumstance (Duncan, 1973). Qlueck (1977) defined decision-making as a process of thought and deliberation which eventually leads to a decision. The preference and choice, and sometimes achievement or lose, success and failure are influenced by our decision. Our personal and professional choices and preferences are reflected in our decision-making. The field of decision-making has evolved over a while, and several disciplines have contributed to it ranging from social to medical sciences. The researcher has presented different models of decision-making based on rationality and involved processing.

An action is called rational if it directly leads to the goal, thus the decision is rational to the extent it leads us to the accomplishment of our goal (Simon, 1993). In rational decision-making, the decision-maker carefully examines the situation to

S. Pathak (✉) · B. L. Srivastava
Department of Humanities and Social Sciences, Indian Institute of Technology Kharagpur, Kharagpur, India
e-mail: kbls@hss.iitkgp.ernet.in

S. N. Mohanty (ed.), *Emotion and Information Processing*,
https://doi.org/10.1007/978-3-030-48849-9_1

maximize the outcome. Whereas irrational decisions are poorly adapted to the goal; though Simon (1993) was doubtful about the existence of excessively rational decisions, he defined irrational decision as one that goes against the logic. Non-rational decisions are intuitive and judgemental; a decision-maker does not sequentially analyse the situation or problem, thus reaches out to a solution in a prompt and fast speed.

1.1 Five Steps in Decision-Making

Decision making involves a five-step process: identification of a situation where decision is needed, developing an alternative action, evaluation of this alternative action, selection and implementation of one selected action (Newman, 1971). A decision-maker does not simply follow those defined steps, but being a person with cognitive as well as emotional abilities attempts to see the whole situation to reach out to a decision which is based on his or her abilities, experiences and capabilities.

1.2 Good or Bad Decision?

Studies suggest that a decision can never be claimed as right or wrong, but the skills and competence involved in this process could be measured to know the construct of the good, skilled and competent decision-making process. The researchers have developed the test to distinguish good, bad and poor decision-making; this test involves measures of different skills and competence required in decision-making (Bruine de Bruin, Parker, & Fischhoff, 2007; Parker & Fischhoff, 2005). The emotions, orientation and motivation of individual, conditions and its background involved variable interplay with each other and influence the competence of decision-maker.

1.3 Act of Thinking in Decision-Making

The act of thinking is very crucial in decision-making as it is the process of critically gathering, analysing and evaluating information. The selection of the final choice among several options is the product of the act of thinking. Some researchers have presented their views on two modes of thinking: system 1 thinking (intuitive) and system 2 thinking (analytical) (Hogarth, 2001; Kahneman & Frederick, 2002; Sloman, 1996). Hammond (1996) considered both of them as not as being distinct but representing the ends of a continuance. Simon (1983) argued about the presence of both analytical and intuitive thoughts in serious and complex thinking. This dual processing view was an extension of an information processing approach of decision task.

Systematic and heuristic information processing describes the receiving and processing of the information. Heuristics processing is characterized by using a generalized rule which is based on experience and observation of the individual. Thus, it provides an economic advantage as minimal cognitive effort is required to derive the overall conclusion. This processing is based on the availability, accessibility and applicability of the given information. In systematic processing, the overall message is based on the reliable sources as it involves comprehensive, cognitive and analytic processing of information (Chaiken, 1980; Chen, Duckworth, & Chaiken, 1999). Unlike heuristic approach, which is only focused on simple rules, the systematic approach follows the detailed processing of information. Chaiken (1980) supports the idea that high involvement leads to systematic information processing in which the persuasion is mediated by message-based cognition, whereas in low involvement simple decision rule mediates persuasion and thus leads to heuristic information processing. In the heuristic process, independent variables such as source factors may influence a recipient's willingness to accept the message conclusion but do not necessarily influence reception or acceptance of argumentation. The conclusion is based on the accessible context information.

Decision style or information processing style is recognized as a subset of cognitive style (Dalal & Brooks, 2013; Kozhevnikov, 2007; Thunholm, 2004). Cognitive and decision style terms have been used interchangeably (Andersen, 2000). Decision style reflects the persistent pattern reflected in decision-making (Driver, 1979). This represents the individual's pattern of perceiving and responding to decision-making situation (Harren, 1979). Decision-maker's cognitive abilities such as information processing, self-evaluation and self-regulation modulate his or her decision style (Thunholm, 2004). Cognitive-Experiential Self-Theory (Epstein, 1994) explains rational and experiential styles. It is also a dual-process theory, but unlike Shelly Chaiken's dual-process model, it is developed in a global theory of personality background. According to this theory, information processing takes place in two modes, rational and experiential modes, and interacts on a holistic level. Rational is normative, systematic work on the conscious level and experiential is unconscious, automatic work based on hunches (Denes-Raj & Epstein, 1994; Epstein, Pacini, Denes-Raj, & Heier, 1996; Kirkpatrick & Epstein, 1992).

1.4 Emotion in Decision-Making

Ekman (1992) defined emotion of an infant is a direct expression of affect. Human emotions are meant to fulfil the social expectation. Affect is a non-conscious experience of intensity; it is an abstract thing which cannot be realized in language, whereas emotions are the projection or display of the feeling. Theorists defined emotion as voice, body movement or innate facial expression. In the nineteenth century, Darwin recognized the innate basis for the facial expression of a certain emotion. Izard (1971) found that the facial expression of primary emotion is innate. Damasio (1994) stated that affect provides intensity without which we can not feel, and in the absence of feeling, rational decision-making becomes problematic.

Simon (1967, 1983) introduced the concept of bounded rationality, redefined the existing rational choice model and included cognitive and situation constraints in the decision-making process. At the same time, researchers become increasingly interested in the role of emotions in the decision-making process. The evolution of somatic marker hypothesis (Damasio, Tranel, & Damasio, 1991) proposed the relationship between a higher brain region and an emotion-based signal arising from the body. This hypothesis was developed on Iowa Gambling Task (Bechara, Tranel, Damasio, Damasio, & Anderson, 1996), an experimental program to replicate real-life decision-making which includes uncertainty, rewards and punishment. They found that damage in prefrontal cortex impairs autonomic response to anticipated future outcome in participants. Damasio (1996) proposed that venrtomedial prefrontal cortex regulates decision-making in a complex and uncertain situation.

1.5 Emotion Mediating the Information Process During the Decision-Making

Emotion as a motivational process Emotion acts as a motivational factor for goal-directed action. Zeelenberg and Pieters (2006) elaborated their perspective "feeling is for doing" rooted in William James pragmatic approach (1907, 1909). Emotion was considered as a motivational process, and the worth of emotion was determined by external criterion like behaviour. Freud (1920/1952) has also stated the true nature of our psyche as "our entire psychical activity is bent upon procuring pleasure and avoiding pain". One specific emotion decides a specific course of action; thus, the specificity of emotions is crucial. Emotion helps to focus on a certain aspect of the given option (Hanoch, 2002). It also acts as a controlling factor for decision-makers to motivate them towards a specific course of action (Frijda, 1986). Researches indicated that two decisions related to emotion, mainly regret and disappointment or guilt or shame, may have a distinct behavioural effect due to the motivational process and individual disposition involved (Zeelenberg, Nelissen, Breugelmans, & Pieters, 2008; Zeelenberg & Pieters, 1999, 2004). Same as anger motivates us to move against the object, fear commits us to move away (Nelissen, Dijker, & de Vries, 2007).

Emotion as an information input Our emotional state also serves as information input during decision-making. Affect as information hypothesis (Clore & Huntsinger, 2007) states that being happy or sad does change both the content and the style of thinking. How we feel about the object of judgement (experiential information) processes faster than objective information. Positive mood is associated with semantic processing, and it leads to semantic arousal and dopamine release (Asbhy, 1999;

Kuhl, 2000). Thus, happy mood leads to relational processing, but sad mood processes information more on item level (Clore & Storbeck, 2006; Isen, 1999).

People evaluate their feeling about the value of judgement object and take feedback from elicited positive or negative emotion (Clore & Martin, 2001). Positive affect trigger relational processing where people relate incoming information to previously known and believed information. Negative affect hinders relational processing; thus, people in sad mood tend to use referential process or item-level processing. The same argument was made in priming hypothesis of memory which states mood primes mood-congruent material in memory (Bower, Monteiro, & Gilligan, 1978).

Evidence suggested that deficit in affect due to neurological damage caused distort judgement and decision-making. Emotions guide our decisions even when we are not aware of it (Keltner & Lerner, 2010). A meta-analytic review which tried to integrate neuroscience experiment of emotion and behaviour and trend showed that emotion acts as an active information input and important predictor in rational behaviour (Xiong, Liu, & Liu, 2014). This study included 19 works of literature and extracts and analysed 30 effect values. This study quoted previous studies which show emotion influences the quality and effectiveness of decision-making (Dolan, 2002; Lerner & Keltner, 2001; Wilson & Schooler, 1991; Zajonc, 1980). The result shows the impact of emotion on rational person behaviour. This study explained the magnetic nuclear resonance perspective of this finding by stating direct association between emotion and prefrontal cortex, nucleus accumbens septum, insular, anterior cingulate, nucleus amygdalae and other brain regions. Here nucleus accumbens septum is responsible for rational person's approach behaviour, and the activation of insular is related to economic man's avoidance behaviour. Positive emotion is associated with overall processing of information, and negative emotion elicits analytical processing of information (Clark & Isen, 1982; Forgas, Bower, & Moylan, 1990; Gasper, 2004).

Integral, incidental and specific emotions in decision-making Lerner, Valdesolo and Kassam (2015) postulated eight major themes related to emotions and decision-making process and stated that emotions powerfully, predictably and persuasively influence the decision-making. They argued that integral emotion, incidental emotions and specific emotions influence decision-making. Judgement and selection on a given task give rise to integral emotions. For example, anxiety resulted because risky choice situation motivates a processing of more safe option rather than a more profitable but risky option. Incidental emotions are carried over emotions which arise as a result of another situation's emotion. These emotions are normatively unrelated to the situation. It elicits a tendency to target a person who is not at all related to the source of anger; we can observe the effect of incidental emotions in our daily lives when we display our frustration of workplace at home. Specific emotions may belong to the same valence but lead to a different effect on decision-makings, such as anger and sadness. Both are of negative valence, but anger activates aggression, whereas fear set-off a flight response. Emotions also shape decision via content and depth of thought processing and goal activation. Thus, emotions influence interpersonal decision-making, for example, negative emotion like anger influences the depth and content of thought processing, as it leads to the attribution of negative events caused by others. Fear involves a low sense of control over the situation, so people tend to attribute situational factors as the cause of a negative event. They also highlighted the possible undesirable effects of emotion on decision-making and its management.

Lowenstein and Lerner (2003) highlighted the role of immediate emotion—an emotion experienced at the time of decision-making—to extend the work of conventional theorist who only included expected emotion in decision-making and discarded the dichotomy between good and bad influence of affect. They stated that both immediate and expected emotions are crucial for decision-making and can lead to decision error too in different ways. The presence of emotion helps to regulate strategies and guide behaviour in more adaptive ways.

Categorical and dimensional emotion model These two models predict the influence of emotion on cognitive processing of information. According to categorical emotion model, primary emotion is six to ten in numbers and not subjected to be reduced in dimensions (Ekman, 1984, 1992; Izard 1977). This model suggests that each emotion is related to a central node which represents a unit of information. Once this unit is triggered due to any physical or psychological stimuli, it reaches a threshold, triggers only related and relevant information node and further evokes characteristic autonomic, expressive behaviour. So, sadness will lead to the processing of only sadness-related information, not anger or fear. This is known as associative network, spreading activation model of memory (Bower, 1981, 1987; Clark & Isen, 1982). Niedenthal and Setterlund (1994) used lexical decision methodology and found emotion congruent word recognition which is consistent with the categorical model. Being in an emotional state leads to the processing of those associated information which influences the interpretation and judgement of an event (Bower, 1981).

Dimensional model suggests that emotions are experienced as dimensional such as valence (Averill, 1975; Osgood, Suci, & Tannenbaum, 1957; Russell, 1980). This model attempts to define emotion in two or three dimensions. Wundt (1897) explained three dimensions of emotion named as "pleasurable versus unpleasurable", "arousing or subduing" and "strain or relaxation". Valence, arousal and intensity are examples of such dimension. Here one common and the coordinated neurophysiological system is considered responsible for every affective states. So, if a person is sad, all related dimensional words such as fear and anger will also be processed.

Social information (EASI) model of emotion Emotion also facilitates social information processing in a certain social situation to remove ambiguity. Van Kleef, De Dreu, and Manstead (2010) presented the social information (EASI) model of emotion which assumes that in a social situation we use emotional cues of others to understand the ambiguous situation. The better understanding of social decision-making was crucial to understanding the decision-making process. Their work suggested that affective reactions take place in a cooperative social situation, and when the epistemic motivation of decision-maker is not very high. Yip and Côté (2013) found that the presence of high emotional intelligence was directly related to accurate identification of the root cause event of emotions, thus helpful in reducing the potential impact of such incidental emotion.

Emotion in dual information processing The presence of positive and negative emotions influences the depth of processing via activating heuristic or systematic information processing in decision-making (Lerner & Tiedens, 2006; Small & Lerner, 2008; Tiedens & Linton, 2001). Tiedens and Linton (2001) investigated and compared the concerned emotion associated with the certainty and uncertainty appraisal on undergraduate students and concluded that certainty appraisal of emotion decides systematic or heuristic processing of the situation. Results suggest that certainty also affects the depth of processing and certainty in one situation perpetuates congruent judgement in the subsequent situation. Responses lead to the conclusion that certainty leads to heuristic processing, whereas uncertainty leads to systematic processing. Weary and Jacobson (1997) have shown that feeling of uncertainty leads to a more systematic process of information. They found that depressed people, who are always uncertain, systematically process the information. Lack of confidence in own judgement was considered the main reason behind this systematic processing (Edwards & Weary, 1993; Gleicher & Weary, 1991; Weary, 1990).

Participants in a negative mood reflect more fluency, originality and flexibility during information processing (Mohanty & Suar, 2014). These components are suggestive of systematic processing. Positive mood influences the information proceeding by proving a safe feeling, and it reduces the need for elaboration and increases confidence, thus leads to general, fast and simple processing. Negative mood state triggers a feeling of threat in the situation; thus, a decision-maker systematically processes the situation.

The review of research work on decision-making indicated the complexities involved in this process; therefore, labelling of a good or bad decision is a complicated task. The assessment of decision-making style, preference and competence provides a better understanding of decision-making as a process. Experiential decision style is found to be related to heuristic processing. Uncertainty and complexity in a situation and individual differences in cognitive style predict the use of heuristic processing (Bröder 2003; Hilbig, 2008; Hogarth & Karelaia, 2006; Newell, Weston, & Shanks, 2003).

Risk perception and judgement in an emotional state Lerner, Gonzalez, Small, and Fischhoff (2003) highlighted the role of positive and negative emotions on risk perception in decision-making. The result indicates that the content of thought, anger and fear determined the high- and low-risk perception in the participants in the wake of 9/11 attacks in the USA. The participants with fear-induced emotion reported less optimistic beliefs than experiencing anger-induced emotion. Anger reduced the risk perception of the situation by triggering intuitive heuristic processing. Fear-related information processes on system level 1 and leads to risk aversion (Lerner & Keltner 2001; Mohanty & Suar, 2013). The fundamental way of risk perception, risk as feeling, reflects our instinctive and intuitive reaction. Fear comprises an appraisal of uncertainty and thus increases risk perception. Whereas anger involves a certainty-related feeling and sense of control and thus reduces risk perception (Slovic & Peters, 2006). Positive or negative affect motivates decision-maker to such action which increases or avoids such feeling, respectively.

Ambady and Gray (2002) studied the effect of mood on judgement among undergraduate student and found that sadness impacts information processing and reduces accuracy in judgement. Students who were induced with sadness with the help of video clips showed poor sensitivity towards other's intention and affect and were not able to reach to the conclusion in subsequent tasks. Introduction of any distraction improves information processing pattern in sad participants and makes them able to understand the intention of others.

The dark side of emotion Information processing in uncertain and certain situations involves different neural network and identifies the dual role of emotion as it may or may not help in the decision-making process (Bechara, 2005; Shiv, Loewenstein, & Bechara, 2005). Shiv et al. (2005) further investigated the possible disadvantageous effect of emotion in comparing two groups of individuals: one was a substance-dependent group of individuals surviving with focal brain lesion in the emotion-related area and another was a normal group with no such lesion. The patients with lesion made a more advantageous decision, and the normal group was more reluctant to invest after winning or losing money on investment round. The role of emotion was proved to be unfavourable in the decision-making of the normal group. Happy people tend to avoid systematic processing and see the quality of argument in a given task. It was argued that happiness reduces cognitive capacity or processing motivation (Isen 1987; Mackie & Worth, 1989). The inspiration to maintain a positive state of mind leads to reduced effort in cognitive processing of available information.

Happy people tend to overlook the quality of argument and indulge in heuristic processing but do perform better on the given task. This summarized that positive state only reduces the motivation to process on the contrary of reducing the whole cognitive capacity. Individuals in a positive mood state see the data they simply don't utilize it (Bless, Bohner, Schwarz, & Strack, 1990; Mohanty & Suar, 2013).

Gender difference in emotional state and information processing The gender difference in decision-making style is evident at early childhood (Crone, Bunge, Latenstein, & van der Molen, 2005; Overman et al., 2004). IOWA gambling task shows gender differences in decision-making. Man tends to choose long-term advantageous options over short-term reward, and women are found to be very sensitive to immediate loss (Bolla, Eldreth, Matochik, & Cadet, 2004; Weller, Levin, & Bechara, 2010). Though other studies presented contradictory findings and reported no gender difference in risk based decision making (Lighthall, Mather, & Gorlick, 2009; Lighthall et al. 2012; Starcke, Wolf, Markowitsch, & Brand, 2008).

The experiential style is dominant among women participants; they tend to use intuition processing in an emotional state (Pacini & Epstein, 1999). Rule learning and feedback processing follow the same pattern for both the genders, but they differ on risk adjustment (Reavis & Overman, 2001). Men prefer global processing, whereas women follow detailed information given at hand (Knyazev, Slobodskoj-Plusnin, & Bocharov, 2010). This in turn is significantly related to the allocation of more resources in information processing in women. Women utilize external information in the decision-making process. They have been identified better at recognizing facial expression (Biele & Grabowska, 2006; Hampson, van Anders, & Mullin, 2006). Men use unconscious and rapid processing in an anger-related situation. Neuroscience studies reported that cognitive control and emotional regulation significantly differ in both the gender which is reflected in decision-making style of male and female decision-makers. Brain development, genetic factor and interaction with environmental factor shape such differences (Butler et al., 2005; Mak, Hu, Zhang, Xiao, & Lee, 2009).

1.6 Conclusion

This chapter illustrates the effect of emotion in information processing for decision-making. A general understanding of decision making involved thought process; emotions have been presented with the relevant research findings. The role and influence of emotion have been recognized as a motivational process. Emotion helps to maintain or avoid a certain aspect of choice and leads to goal-directed action. Emotion also works as input information in decision-making. Our feeling regarding the object raises and guides the information processing, and also positive and negative mood states lead to relational or item-level processing. Neurochemical changes and brain region activation generated due to emotional state are crucial in this processing. The characteristic integral, incidental and immediate emotion influences information processing in decision-making. The categorical and dimensional model explains information processing as based upon the associative network, spreading activation and valence, respectively. Social information (EASI) model of emotion explains the use of social cues in understanding and processing social information. Positive emotion state facilitates heuristic and negative emotion state facilitates systematic processing. Emotional states such as anger, fear and sadness manipulate the risk perception and judgement. Happy people tend to overlook the quality of argument and indulge in heuristic processing but do perform better on the given task. Gender difference in decision-making do exist in terms of information processing at decision style, global and specific level of processing, but both genders show the same pattern in rule learning, safe choice and feedback processing.

References

Ambady, N., & Gray, H. M. (2002). On being sad and mistaken: Mood effects on the accuracy of thin-slice judgments. *Journal of Personality and Social Psychology, 83*(4), 947.

Andersen, J. A. (2000). Intuition in managers. *Journal of Managerial Psychology, 15*(1), 46–63.

Averill, J. R. (1975). *A semantic atlas of emotional concepts* (American Psychological Association, Journal Supplement Abstract Service). Washington, DC: American Psychological Association.

Bechara, A. (2005). Decision making, impulse control and loss of willpower to resist drugs: A neurocognitive perspective. *Nature Neuroscience, 8*(11), 1458–1463.

Bechara, A., Tranel, D., Damasio, H., & Damasio, A. R. (1996). Failure to respond autonomically to anticipated future outcomes following damage to prefrontal cortex. *Cerebral Cortex, 6*(2), 215–225.

Biele, C., & Grabowska, A. (2006). Sex differences in perception of emotion intensity in dynamic and static facial expressions. *Experimental Brain Research, 171*(1), 1–6.

Bless, H., Bohner, G., Schwarz, N., & Strack, F. (1990). Mood and persuasion: A cognitive response analysis. *Personality and Social Psychology Bulletin, 16*, 331–345.

Bolla, K. I., Eldreth, D. A., Matochik, J. A., & Cadet, J. L. (2004). Sex-related differences in a gambling task and its neurological correlates. *Cerebral Cortex, 14*(11), 1226–1232.

Bower, G. H. (1981). Mood and memory. *American Psychologist, 36*(2), 129.

Bower, G. H. (1987). Commentary on mood and memory. *Behaviour Research and Therapy, 25*(6), 443–455.

Bower, G. H., Monteiro, K. P., & Gilligan, S. G. (1978). Emotional mood as a context for learning and recall. *Journal of Verbal Learning and Verbal Behavior, 17*(5), 573–585.

Bröder, A. (2003). Decision making with the "adaptive toolbox": Influence of environmental structure, intelligence, and working memory load. *Journal of Experimental Psychology: Learning, Memory, and Cognition, 29*(4), 611–625.

Bruine de Bruin, W., Parker, A. M., & Fischhoff, B. (2007). Individual differences in adult decision-making competence. *Journal of Personality and Social Psychology, 92*(5), 938.

Butler, T., Pan, H., Epstein, J., Protopopescu, X., Tuescher, O., Goldstein, M., ... Ledoux, J. (2005). Fear-related activity in subgenual anterior cingulate differs between men and women. *Neuroreport, 16*(11), 1233–1236.

Cahill, L., & McGaugh, J. L. (1998). Mechanisms of emotional arousal and lasting declarative memory. *Trends in Neurosciences, 21*(7), 294–299. https://pubmed.ncbi.nlm.nih.gov/9683321/

Carpenter, S. K., & DeLosh, E. (2005). Application of the testing and spacing effects to name learning. *Applied Cognitive Psychology, 19*(5), 619–636. https://doi.org/10.1002/acp.1101

Chaiken, S. (1980). Heuristic versus systematic information processing and the use of source versus message cues in persuasion. *Journal of Personality & Social Psychology, 39*(5), 752–766.

Chen, S., Duckworth, K., & Chaiken, S. (1999). Motivated heuristic and systematic processing. *Psychological Inquiry, 10*(1), 44–49.

Clark, M. S., & Isen, A. M. (1982). Toward understanding the relationship between feeling states and social behavior. *Cognitive Social Psychology, 73*, 108.

Clore, G. L., & Huntsinger, J. R. (2007). How emotions inform judgment and regulate thought. *Trends in Cognitive Sciences, 11*(9), 393–399.

Clore, G. L., & Martin, L. L. (2001). *Theories of mood and cognition: A user's guidebook.* Mahwah, NJ: Erlbaum.

Clore, G. L., & Storbeck, J. (2006). *Affect as information about liking, efficacy, and importance.* East Sussex, England: Psychology Press.

Crone, E. A., Bunge, S. A., Latenstein, H., & van der Molen, M. W. (2005). Characterization of children's decision making: sensitivity to punishment frequency, not task complexity. *Child Neuropsychology, 11*(3), 245–263.

Dalal, R. S., & Brooks, M. E. (2013). Individual differences in decision-making skill and style. In *Judgment and decision making at work* (pp. 100–121). Abingdon, UK: Routledge.

Damasio, A. R. (1994). *Descartes error: Emotion, reason and the human brain* (pp. 350–412). New York, NY: Avon.

Damasio, A. R. (1996). The somatic marker hypothesis and the possible functions of the prefrontal cortex. *Philosophical Transactions of the Royal Society of London. Series B: Biological Sciences, 351*(1346), 1413–1420.

Damasio, A. R., Tranel, D., & Damasio, H. C. (1991). Somatic markers and the guidance of Behavior: Theory and preliminary testing. In H. S. Levin, H. M. Eisenberg, & A. L. Benton (Eds.), *Frontal lobe function and dysfunction* (pp. 217–229). New York, NY: Oxford University Press.

Denes-Raj, V., & Epstein, S. (1994). Conflict between intuitive and rational processing: When people behave against their better judgment. *Journal of Personality and Social Psychology, 66*(5), 819.

Dolan, R. J. (2002). Emotion, cognition, and behavior. *Science, 298*(5596), 1191–1194.

Driver, M. J. (1979). Individual decision-making and creativity. In S. Kerr (Ed.), *Organizational behavior*. Columbus, OH: Grid Publishing.

Duncan, W. J. (1973). *Decision making and social issues: A guide to administrative action in an environmental context*. Hinsdale, IL: Dryden Press.

Edwards, J. A., & Weary, G. (1993). Depression and the impression-formation continuum: Piecemeal processing despite the availability of category information. *Journal of Personality and Social Psychology, 64*(4), 636.

Ekman, P. (1984). Expression and the nature of emotion. *Approaches to Emotion, 3*(19), 344.

Ekman, P. (1992). An argument for basic emotions. *Cognition & Emotion, 6*(3–4), 169–200.

Epstein, S. (1994). Integration of the cognitive and the psychodynamic unconscious. *American Psychologist, 49*(8), 709.

Epstein, S., Pacini, R., Denes-Raj, V., & Heier, H. (1996). Individual differences in intuitive–experiential and analytical–rational thinking styles. *Journal of Personality and Social Psychology, 71*(2), 390.

Forgas, J. P., Bower, G. H., & Moylan, S. J. (1990). Praise or blame? Affective influences on attributions for achievement. *Journal of Personality and Social Psychology, 59*(4), 809.

Freud, S. (1920/1952). *A general introduction to psychoanalysis*. New York, NY: Washington Square Press.

Frijda, N. H. (1986). *The emotions*. Cambridge, UK: Cambridge University Press.

Gasper, K. (2004). Do you see what I see? Affect and visual information processing. *Cognition and Emotion, 18*(3), 405–421.

Gleicher, F., & Weary, G. (1991). Effect of depression on quantity and quality of social inferences. *Journal of Personality and Social Psychology, 61*(1), 105.

Hamann, S. (2001). Cognitive and neural mechanisms of emotional memory. *Trends in Cognitive Sciences, 5*(9), 394–400. https://www.researchgate.net/publication/11824502_Cognitive_and_neural_mechanisms_of_emotional_memory

Hammond, K. R. (1996). *Human judgment and social policy: Irreducible uncertainty, inevitable error, unavoidable injustice*. New York, NY: Oxford University Press.

Hampson, E., van Anders, S. M., & Mullin, L. I. (2006). A female advantage in the recognition of emotional facial expressions: Test of an evolutionary hypothesis. *Evolution and Human Behavior, 27*(6), 401–416.

Hanoch, Y. (2002). "Neither an angel nor an ant": Emotion as an aid to bounded rationality. *Journal of Economic Psychology, 23*(1), 1–25.

Harren, V. A. (1979). Research with the assessment of career decision making. *Character Potential: A Record of Research, 9*(2), 63–69.

Hilbig, B. E. (2008). Individual differences in fast-and-frugal decision making: Neuroticism and the recognition heuristic. *Journal of Research in Personality, 42*(6), 1641–1645.

Hogarth, R. M. (2001). *Educating intuition*. Chicago, IL: University of Chicago Press.

Hogarth, R. M., & Karelaia, N. (2006). "Take-the-best" and other simple strategies: Why and when they work "well" with binary cues. *Theory and Decision, 61*(3), 205–249.

Isen, A. M. (1987). Positive affect, cognitive processes, and social behavior. In L. Berkowitz (Ed.), *Advances in experimental social psychology* (Vol. 20, pp. 203–253). New York, NY: Academic Press.

Isen, A. M. (1999). On the relationship between affect and creative problem solving. *Affect, Creative Experience, and Psychological Adjustment, 3*, 17.
Izard, C. E. (1971). *The face of emotion*. New York, NY: Appleton century –Crofts.
Izard, C. E. (1977). *Human emotions*. New York, NY: Plenum.
James, W. (1907/1995). *Pragmatism*. New York, NY: Dover Publications.
James, W. (1909/2002). *The meaning of truth*. New York, NY: Dover Publications.
Kahneman, D., & Frederick, S. (2002). Representativeness revisited: Attribute substitution in intuitive judgment. *Heuristics and Biases: The Psychology of Intuitive Judgment, 49*, 81.
Keltner, D., & Lerner, J. S. (2010). Emotion. In D. T. Gilbert, S. T. Fiske, & G. Lindzey (Eds.), *The handbook of social psychology* (pp. 317–352). New York, NY: Wiley.
Kirkpatrick, L. A., & Epstein, S. (1992). Cognitive-experiential self-theory and subjective probability: Further evidence for two conceptual systems. *Journal of Personality and Social Psychology, 63*(4), 534.
Knyazev, G. G., Slobodskoj-Plusnin, J. Y., & Bocharov, A. V. (2010). Gender differences in implicit and explicit processing of emotional facial expressions as revealed by event-related theta synchronization. *Emotion, 10*(5), 678.
Kozhevnikov, M. (2007). Cognitive styles in the context of modern psychology: Toward an integrated framework of cognitive style. *Psychological Bulletin, 133*(3), 464.
Kuhl, J. (2000). A functional-design approach to motivation and self-regulation: The dynamics of personality systems interactions. In M. Boekaerts, P. R. Pintrich, & M. Zeidner (Eds.), *Handbook of self-regulation* (pp. 111–169). San Diego, CA: Academic Press.
Lerner, J. S., Gonzalez, R. M., Small, D. A., & Fischhoff, B. (2003). Effects of fear and anger on perceived risks of terrorism: A national field experiment. *Psychological Science, 14*(2), 144–150.
Lerner, J. S., & Keltner, D. (2001). Fear, anger, and risk. *Journal of Personality and Social Psychology, 81*(1), 146–159.
Lerner, J. S., Li, Y., Valdesolo, P., & Kassam, K. S. (2015). Emotion and decision making. *Annual Review of Psychology, 66*, 799–823.
Lerner, J. S., & Tiedens, L. Z. (2006). Portrait of the angry decision maker: How appraisal tendencies shape anger's influence on cognition. *Journal of Behavioral Decision Making, 19*(2), 115–137.
Leutner, D. (2014). Motivation and emotion as mediators in multimedia learning. *Learning and Instruction, 29*, 174–175. https://www.researchgate.net/publication/259137983_Motivation_and_emotion_as_mediators_in_multimedia_learning
Lighthall, N. R., Mather, M., & Gorlick, M. A. (2009). Acute stress increases sex differences in risk seeking in the balloon analogue risk task. *PLoS One, 4*(7).
Lighthall, N. R., Sakaki, M., Vasunilashorn, S., Nga, L., Somayajula, S., Chen, E. Y., … Mather, M. (2012). Gender differences in reward-related decision processing under stress. *Social Cognitive and Affective Neuroscience, 7*(4), 476–484.
Mackie, D. M., & Worth, L. T. (1989). Processing deficits and the mediation of positive affect in persuasion. *Journal of Personality and Social Psychology, 57*(1), 27.
Mak, A. K., Hu, Z. G., Zhang, J. X., Xiao, Z., & Lee, T. M. (2009). Sex-related differences in neural activity during emotion regulation. *Neuropsychologia, 47*(13), 2900–2908.
Malone, T. W. (1981). Toward a theory of intrinsically motivating instruction. *Cognitive Science, 5*(4), 333–369. https://doi.org/10.1016/S0364-0213(81)80017-1
Mohanty, S. N., & Suar, D. (2013). Influence of mood states, group discussion, and interpersonal comparison on change in decision-making and information processing. *International Journal of Decision Science, Risk and Management, 5(2)*, 101–123.
Mohanty, S. N., & Suar, D. (2014). Decision making under uncertainty and information processing in positive and negative mood states. *Psychological Reports, 115*(1), 91–105.
Nelissen, R. M., Dijker, A. J., & de Vries, N. K. (2007). Emotions and goals: Assessing relations between values and emotions. *Cognition and Emotion, 21*(4), 902–911.

Newell, B. R., Weston, N. J., & Shanks, D. R. (2003). Empirical tests of a fast-and-frugal heuristic: Not everyone "takes-the-best". *Organizational Behavior and Human Decision Processes, 91*(1), 82–96.
Newman, J. W. (1971). *Management applications of decision theory*. New York, NY: Harper and Row.
Niedenthal, P. M., & Setterlund, M. B. (1994). Emotion congruence in perception. *Personality and Social Psychology Bulletin, 20*(4), 401–411.
Osgood, C. E., Suci, G. J., & Tannenbaum, P. H. (1957). *The measurement of meaning*. Urbana, IL: University of Illinois Press.
Overman, W. H., Frassrand, K., Ansel, S., Trawalter, S., Bies, B., & Redmond, A. (2004). Performance on the IOWA card task by adolescents and adults. *Neuropsychologia, 42*(13), 1838–1851.
Pacini, R., & Epstein, S. (1999). The relation of rational and experiential information processing styles to personality, basic beliefs, and the ratio-bias phenomenon. *Journal of Personality and Social Psychology, 76*(6), 972.
Parker, A. M., & Fischhoff, B. (2005). Decision-making competence: External validation through an individual-differences approach. *Journal of Behavioral Decision Making, 18*(1), 1–27.
Qlueck, W. F. (1977). *Management*. Hinsdale, IL: Dryden Press.
Reavis, R., & Overman, W. H. (2001). Adult sex differences on a decision-making task previously shown to depend on the orbital prefrontal cortex. *Behavioral Neuroscience, 115*(1), 196.
Russell, J. A. (1980). A circumplex model of affect. *Journal of Personality and Social Psychology, 39*(6), 1161.
Shiv, B., Loewenstein, G., & Bechara, A. (2005). The dark side of emotion in decision-making: When individuals with decreased emotional reactions make more advantageous decisions. *Cognitive Brain Research, 23*(1), 85–92.
Simon, H. A. (1967). Motivational and emotional controls of cognition. *Psychological Review, 74*(1), 29.
Simon, H. A. (1983). *Reason in human affairs*. Stanford, CA: Stanford University Press.
Simon, H. A. (1993). Decision making: Rational, nonrational and irrational. *Educational Administrative Quarterly, 29*(3), 392–411.
Sloman, S. A. (1996). The empirical case for two systems of reasoning. *Psychological Bulletin, 119*(1), 3.
Slovic, P., & Peters, E. (2006). Risk perception and affect. *Current Directions in Psychological Science, 15*(6), 322–325.
Small, D. A., & Lerner, J. S. (2008). Emotional policy: Personal sadness and anger shape judgments about a welfare case. *Political Psychology, 29*(2), 149–168.
Starcke, K., Wolf, O. T., Markowitsch, H. J., & Brand, M. (2008). Anticipatory stress influences decision making under explicit risk conditions. *Behavioral Neuroscience, 122*(6), 1352.
Thunholm, P. (2004). Decision-making style: Habit, style or both? *Personality and Individual Differences, 36*(4), 931–944.
Tiedens, L. Z., & Linton, S. (2001). Judgment under emotional certainty and uncertainty: the effects of specific emotions on information processing. *Journal of Personality and Social Psychology, 81*(6), 973.
Van Kleef, G. A., De Dreu, C. K., & Manstead, A. S. (2010). An interpersonal approach to emotion in social decision making: The emotions as social information model. In *Advances in experimental social psychology* (Vol. 42, pp. 45–96). New York, NY: Academic Press.
Weary, G. (1990). Depression and sensitivity to social information. In A. M. Isen & B. S. Moore (Eds.), *Affect and social behavior* (pp. 207–230). New York, NY: Cambridge University Press.
Weary, G., & Jacobson, J. A. (1997). Causal uncertainty beliefs and diagnostic information seeking. *Journal of Personality and Social Psychology, 73*(4), 839.
Weller, J. A., Levin, I. P., & Bechara, A. (2010). Do individual differences in Iowa Gambling Task performance predict adaptive decision making for risky gains and losses? *Journal of Clinical and Experimental Neuropsychology, 32*(2), 141–150.

Wilson, T. D., & Schooler, J. W. (1991). Thinking too much: Introspection can reduce the quality of preferences and decisions. *Journal of Personality and Social Psychology, 60*(2), 181.
Wundt, W. M. (1897). Outlines of psychology. In *Classics in the history of psychology*. http://psychclassics.asu.edu/index.htm Archived 2001-02-24 at the Wayback Machine, York University 2010, Toronto.
Xiong, G., Liu, Y., & Liu, X. (2014). The meta-analysis on the effect of emotion on rational person behavior in the decision neuroscience perspective. *The Open Cybernetics &Systemics Journal, 8*(1), 1123–1128.
Yip, J. A., & Côté, S. (2013). The emotionally intelligent decision maker: Emotion-understanding ability reduces the effect of incidental anxiety on risk taking. *Psychological Science, 24*(1), 48–55.
Zajonc, R. B. (1980). Feeling and thinking: Preferences need no inferences. *American Psychologist, 35*(2), 151.
Zeelenberg, M., Nelissen, R. M., Breugelmans, S. M., & Pieters, R. (2008). On emotion specificity in decision making: Why feeling is for doing. *Judgment and Decision making, 3*(1), 18.
Zeelenberg, M., & Pieters, R. (1999). Comparing service delivery to what might have been: Behavioral responses to regret and disappointment. *Journal of Service Research, 2*(1), 86–97.
Zeelenberg, M., & Pieters, R. (2004). Beyond valence in customer dissatisfaction: A review and new findings on behavioral responses to regret and disappointment in failed services. *Journal of Business Research, 57*(4), 445–455.
Zeelenberg, M., & Pieters, R. (2006). Feeling is for doing: A pragmatic approach to the study of emotions in economic behavior. In D. De Cremer, M. Zeelenberg, & J. K. Murnighan (Eds.), Social psychology and economics (pp. 117–137). Mahwah, NJ: Erlbaum

Chapter 2
Impact of Psychosocial Intervention on Emotional Disorders and Medication Adherence in HIV/AIDS

Rajeev Kumar, Damodar Suar, and Sanjay Kumar Singh

Contents

2.1 Introduction

The status of being HIV positive evokes a grievous reaction and death-related anxiety (Hintze, Templer, Cappelletty, & Frederick, 1993). Moreover, exclusion from the family and society and the discrimination perpetuate those negative emotional reactions (Attrill, Kinniburgh, & Power, 2001; Chan, Stoové, & Reidpath, 2008). This psychological distress and emotional disorders cause deterioration in the immune system, and the downfall of CD4 counts (Goforth, Lowery, Cutson, Kenedi, & Cohen, 2009). Depression and anxiety are ubiquitous emotional disorders among people living with HIV/AIDS (PLHA) (Olley, Seedat, & Stein, 2006; Primeau, Avellaneda, Musselman, St. Jean, & Illa, 2013). Other symptoms among PLHA are posttraumatic stress disorder (Breet, Kagee, & Seedat, 2014), suicide attempts, suicidal ideation (Catalan et al., 2011), alcohol consumption (Kumar, Suar, Singh, & Bhattaharya, 2019), psychotic disorders(Nebhinani & Mattoo, 2013), and neurocognitive disorders (Robertson et al., 2014).

R. Kumar (✉) · D. Suar
Indian Institute of Technology Kharagpur, Kharagpur, West Bengal, India
e-mail: ds@hss.iitkgp.ernet.in

S. K. Singh
Rajendra Institute of Medical Sciences, Ranchi, Jharkhand, India

S. N. Mohanty (ed.), *Emotion and Information Processing*,
https://doi.org/10.1007/978-3-030-48849-9_2

Psychosocial interventions with antiretroviral therapies are warranted for improving the efficacy of treatment in HIV/AIDS. Cognitive-behavioral therapy is found effective in treating depression and anxiety disorders in HIV/AIDS (Jayasvasti et al., 2011; Spies, Asmal, & Seedat, 2013). Motivational enhancement therapy has yielded fruitful outcomes in reducing relapses and maintaining abstinence in the substance-use problem among PLHA (Durvasula & Miller, 2014). Peer support (Kumakech, Cantor-Graae, Maling, & Bajunirwe, 2009) and supportive therapy (Maldonado et al., 1996) imbibe a sense of acceptance, reduce psychological distress, and enhance the quality of life (Hosseinian, Tabatabaei, & Ravaei, 2011). In psychosocial interventions, cultural sensitivities need to be taken care of while handling individual issues (Wicks, 1996). Earlier literature on HIV/AIDS has shown the paucity of interventional studies. HIV/AIDS encompasses a wide range of psychosocial problems, which require holistic intervention for greater efficacy. The social diagnosis of HIV presents multi-layers of the issue; each layer is connected with another. Those layers include a barrier to healthcare access, discrimination, social exclusion, psychological distress, a misconception of HIV, lifestyle, and nutrition. All those multi-layer issues ultimately affect the prognosis of HIV/AIDS (Fig. 2.1). Therefore, this study assesses whether the integrated psychosocial intervention among PLHA can improve their psychosocial and clinical conditions.

2.2 Method

2.2.1 Design

This study employed the pre-post experimental design. The study examines the effects of psychosocial intervention on emotional disorders and social issues faced by persons infected with HIV/AIDS. This study was conducted at the antiretroviral therapy (ART) center of Jharkhand (India). Ten PLHA having psychosocial problems were selected conveniently, who were willing to participate in the study, and signed an informed consent form. However, only six participants completed all sessions for psychosocial intervention spanning 6 months.

2.2.2 Ethical Concern

The Institute Ethics Committees of Indian Institute of Technology Kharagpur and Rajendra Institute of Medical Sciences, Ranchi, have approved this study.

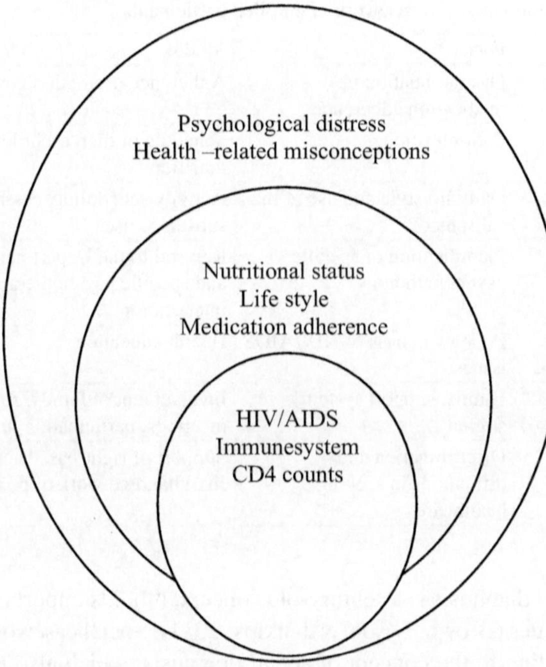

Fig. 2.1 Social diagnosis of HIV/AIDS

2.2.3 Procedure

After interacting with all cases, it was observed that most of the people enrolled in ART center face common psychosocial issues: medication adherence, nutritional problem, various psychological and emotional distress, social exclusion, low-income family support, perceived stigma, and discrimination at multiple places. The technique of social casework was adopted to address the issues participants faced. While social casework addresses a wide range of psychosocial issues holistically, on the other hand, single psychotherapy focuses on psychiatric symptoms only (Mathew, 1992). Social casework involved a multidisciplinary approach and utilization of available resources. The individual casework of all enrolled participants was done. Their psychiatric diagnosis was finalized after consultation with the Department of Psychiatry, Rajendra Institute of Medical Sciences, Ranchi. A psychiatric diagnosis was made according to the diagnostic guidelines of the International Classification of Diseases-10 (World-Health-Organization, 1992).

Table 2.1 Formulation of social casework of enrolled participants

Locus	Focus	Modus
Medication adherence	Discontinuation in medication adherence	Adherence promotion counseling
Nutrition	Poor nutrition	Schedule of diet, according to available nutrition
Lifestyle	Daily lifestyle and use of the substance	Activity scheduling, cessation of substance use
Psychological distress	Identification of specific psychopathology	Referral to the Department of Psychiatry and specific psychotherapeutic intervention
Health related misconceptions	Poor awareness of HIV/AIDS issues	Health education
Social exclusion	Family, support system, stigma	Involvement of family, support group meetings, participation in voluntary work
Healthcare barrier	Discrimination and difficulties in seeking healthcare	Support of right-based activism of Jharkhand network of positive people

Though social diagnosis is a century-old concept, it holds importance in contemporary health issues (Brown, Lyson, & Jenkins, 2011). Social casework formulation was done according to the concept of social diagnosis, which also includes social determinants of health, such as nutrition and stigma (Richmond, 1917). The counselor of the ART center conducted adherence promotion counseling. A nurse of the ART center undertook nutritional counseling; accordingly, the dietary schedule was developed, which was appropriate to the culture and available resources. Activity scheduling was done for all cases, and they were encouraged to follow it because good medication adherence depend on a healthy lifestyle. All cases were discouraged for the intake of alcohol and tobacco. To resolve misconceptions related to HIV/AIDS, health education was imparted. Family members, especially spouses, regardless of their seroconversion status, participated in this intervention package. Jharkhand Network of Positive People cooperated for conducting support group meetings, combating discrimination, and facilitating help needed for HIV-infected people. Jharkhand Network of Positive People is a support group and community service center authorized by National AIDS Control Organization, New Delhi (Table 2.1).

2.2.4 *Measures*

In five different cases, six PLHA participated in this study. Medication adherence was assessed using pill count method (World Health Organization, 2004):

$$\text{Medication adherence}(\%) = \frac{\text{Number of pills taken}}{\text{number of pills prescribed}} \times 100.$$

CD4+ T-cell counts were examined at an interval of 6 months as per official norms (World-Health-Organization, 2004). To measure the severity of hypochondriacal symptoms, Whitely Index-7 was used, and responses to seven items were measured on a 5-point Likert scale. A higher score of the scale showed higher symptoms of hypochondriasis; its validity and Cronbach's alpha of 0.60 were established earlier (Fink et al., 1999). Hamilton Anxiety Scale measured 14 symptoms of anxiety on a 5-point Likert scale from 0 (not present) to 4 (severe); its Cronbach's alpha of 0.80 was reported earlier (Hamilton, 1959). Marital conflicts were assessed using the Marital Communication Inventory having 45 items; responses of the scale were obtained on a 4-point Likert scale from 1 (usually) to 4 (never) for the couple under psychotherapy. This scale had high reliability of 0.90 and good validity (Bienvenu, 1970); its higher score reflected the severity of marital communication. Beck Depression Inventory measured the intensity of 21 depressive symptoms on a 4-point Likert scale from 0 (not present) to 3 (severe); a higher score indicated severity of depressive symptomatology. It had high-reliability coefficients of 0.76–0.95 (Beck, Steer, & Carbin, 1988). Using Temptation and Restraint Inventory, preoccupation with alcohol consumption was rated, inventory had 15 items, and its responses were given on a 9-point (1–9) Likert scale, higher score expressed an increased preoccupation with alcohol consumption. This inventory had high reliability of 0.78–0.91 and acceptable validity (Collins & Lapp, 1992).

2.2.5 *Statistical Analysis*

Medication adherence and CD4+ T-cell counts were the common outcome variables for the participants. Since CD4+ T-cell counts were monitored only twice during therapeutic sessions, and medication adherence at the endline was noted similar to the midline. Therefore, paired-sample "*t*" test was applied to examine the difference in medication adherence and CD4+ T-cell counts from baseline and endline. The rest of the psychiatric and social issues were monitored using different scales. Their scores of baseline, midline, and endline were compared for each PLHA.

2.3 Result

2.3.1 *Clinical Descriptions*

2.3.1.1 Case 1

The client was a 32-year-old married male, educated up to B.A., residing in the urban area of Ranchi and running his own business. From the last 8 months, he has been suspecting cancer, cardiac disease, and high blood pressure, while on clinical investigation, all suspected problems were ruled out. His lab investigations and distress were impairing his daily activities and medication adherence. He was diagnosed as hypochondriasis, and cognitive behavioral therapy with Jacobson's progressive muscular relaxation techniques was applied to him for 20 weeks.

2.3.1.2 Case 2

The client was 34-year-old, married, educated up to 12 standards, working as a salesman, and reported his symptoms for 4 months that included persistent fear of dying, muscle tension, difficulty in concentration, light-headedness, and dry mouth. His symptoms increased during hospital visit and sight of death. These symptoms were hampering his daily work and medication adherence. He was diagnosed with having a generalized anxiety disorder. It was planned to administer cognitive behavioral therapy, systematic desensitization, and Jacobson's muscle relaxation techniques for 21 weeks.

2.3.1.3 Cases 3 and 4

The clients were 30-year-old male and his 28-year-old wife; they were a seroconcordant couple. He reported frequent marital quarrel after disclosure of his HIV-positive status to his spouse. Their marital conflict hampered their regular follow-ups and medication adherence. After the case formulation, the HIV transmission to his wife was the cause of matrimonial quarrels, which resulted in poor communication and misunderstanding between them. Interpersonal therapy of 12 weeks was planned and offered that decreased their marital conflicts and enhanced adjustment with one another.

2.3.1.4 Case 5

The client was 27-year-old unmarried male, educated up to intermediate, and working in a private company. He reported persistent sad mood, suicidal ideation, one suicide attempt, diminished appetite, and guilt feeling. He was diagnosed with having moderate depression with somatic syndrome. Cognitive behavior therapy was planned and executed for 24 weeks.

2.3.1.5 Case 6

The client was a 44-year-old male, widower, educated up to matriculation, and working as a motor mechanic. His wife had died of AIDS, 2 years back, after that, he started consuming alcohol. He was diagnosed with alcohol-dependence syndrome. Before coming to the ART center, he was discharged from a de-addiction center after detoxification. Motivation enhancement therapy and relapse prevention

Table 2.2 Monitoring of clinical progress

Cases (Monitoring tools)	Baseline score	Midline score	Endline score
Case 1, Hamilton Anxiety Scale	24	17	5
Case 1, Whitely-7 Score	29	15	9
Case 2, Hamilton Anxiety Scale	48	26	11
Cases 3 and 4, Marital Communication	90 (Wife)	50 (Wife)	20 (Wife)
Inventory	70 (Husband)	40 (Husband)	18 (Husband)
Case 5, Beck Depression Inventory	35	17	5
Case 6, Temptation and Restraint Inventory	90	50	10

Table 2.3 Comparison of clinical progress

Clinical outcome	Pre-intervention *M (SD)*	Post-intervention *M(SD)*	*t*	*df*	*p*
CD 4 counts	251.67 (83.85)	517.83 (99.51)	18.55	5	0.0001
Medication adherence	81.83 (6.97)	100 (0)	6.38	5	0.0014

Table 2.4 Psychotherapeutic case formulation

Cases with ICD-10 codes	Locus	Focus	Modus
Case 1 (F 45.2)	Hypochondriasis	Excessive health-related worries, frequent doctor-shopping	Jacobson muscular relaxation techniques, cognitive behavioral therapy, through monitoring, pi-chart method, identification of unproductive worries
Case 2 (F 41.1)	Generalized anxiety disorder	Anxiety related to a hospital visit	Jacobson muscular relaxation techniques, systematic desensitization, challenging automatic thoughts
Cases 3, 4 (Z 63.0)	Marital conflict	Marital discordance affecting treatment adherence	Mutual forgiveness, avoidance of provocative topics, role reversal technique, thought and communication monitoring
Case 5 (F 3, 2.1)	The moderate depressive episode with the somatic syndrome	Guilt feelings suicidal ideation	Cognitive-behavioral therapy, identification of self-defeating thoughts, substitution with positive thoughts, emotional ventilation, suicide prevention techniques
Case 6 (F 10.10)	Alcohol-dependence syndrome	Pessimistic view toward life, loneliness	Motivation enhancement therapy, involvement in voluntary help to people infected with HIV, relapse prevention

techniques were planned to prolong his abstinence. He completed 21 sessions of the designed intervention.

In all cases, their presented symptoms were improved (Table 2.2), CD 4 count, and medication adherence were enhanced (Table 2.3). The data are shown in Tables 2.2 and 2.4, which are self-explanatory.

2.4 Discussion

Enrolled PLHA on ART avoid a prolonged stay in the ART center because of the attached stigma. Many PLHA do not consent for attending sessions of psychosocial intervention. Many of the PLHA come from far-off places. Consequently, it was less feasible for them to do weekly follow-up and participate in hour-long sessions of therapies. All PLHA who consented to participate in the study belonged to local places. Social casework formulation of all cases was chosen for broader coverage and understanding of issues at depth and inclusion of single psychotherapy (Mathew, 1992). The concept of social diagnosis (Richmond, 1917) emphasizes on probing in multiple layers of physical, psychological, social, infrastructural, and all aspects related to the life of persons (Brown et al., 2011). HIV/AIDS is also associated with socio-economic conditions (Kumar, Suar, & Singh, 2017), and the biopsychosocial model is a more extensive explanation of social diagnosis (Engel, 1977).

Case 1 was economic well-off, businessman; he was willing to participate in psychotherapy sessions along with on-going ART. His cooperation in therapy helped in its successful outcome. Earlier research has confirmed the efficacy of cognitive-behavioral therapy in hypochondriasis. Because it is associated with long-term benefits and low dropout rate, it is more effective than any other treatment in hypochondriasis (Bouman & Visser, 1998; Thomson & Page, 2007) and generalized anxiety disorder (Cuijpers et al., 2014). Therefore, cognitive behavioral therapy was the most suitable therapeutic option for this case.

Case 2 was an educated person; therefore, it was feasible to administer sessions of cognitive-behavioral therapy on him. Cognitive-behavioral therapy is effective in treating anxiety disorders (Cuijpers et al., 2014; Jayasvasti et al., 2011; Spies et al., 2013).

In Cases 3 and 4, a blame game had started between the husband and the wife, when the wife was detected HIV positive. It was apparent that the route of HIV transmission was from her husband. Evidence suggests that the majority of HIV-infected women received the virus through their husbands (Deering et al., 2008; Halli, Blanchard, Satihal, & Moses, 2007; Saggurti et al., 2011). Of the HIV-infected women, only 10% are involved in commercial sex work, and the rest of them are monogamous (UNDP PFI, 2011). According to the partner-notification policy, it is mandatory to disclose the positive status of HIV to one's spouse and HIV test of another marital partner (National-AIDS-Control-Organization, 2007). Marital conflicts aggravate among several couples after partner notification of HIV status (Shamu, Zarowsky, Shefer, Temmerman, & Abrahams, 2014). In Indian societies, less educated or illiterate women remain submissive even after partner notification of HIV status because of their ignorance; on the other hand, studies conducted elsewhere have reported a high number of marital separation among HIV-positive couples (Mackelprang et al., 2014). Collectivist societal norms discourage marital dissolution in India. It is seen in Case 3, where his wife was not willing for marital separation, and it is the strength of this case for psychosocial intervention.

Case 5 had guilt feeling of being HIV-infected and made one suicide attempt. Earlier literature has observed suicidal ideation and suicide attempts among PLHA on moral grounds. HIV-infected people are alienated from the social mainstream, and it further pushes them to end their lives (Catalan et al., 2011). Effective results of cognitive-behavioral therapies are found in mild to moderate depression (Cuijpers, van Straten, Andersson, & van Oppen, 2008). Therefore, cognitive behavioral therapy for this case was chosen.

As in Case 6, substance abuse, especially alcohol consumption, is widespread among PLHA (Kumar et al., 2019). Even many cases indulge in high-risk activities and get infected with HIV under the intoxication of substance abuse. Moreover, alcohol consumption causes early AIDS-related fatalities because it weakens the immune system (Kumar et al., 2019). The efficacy of motivational enhancement therapy is found effective among PLHA with alcohol-dependence syndrome and in the prevention of relapses (Durvasula & Miller, 2014) and dropout from therapeutic sessions (Sellman, Sullivan, Dore, Adamson, & MacEwan, 2001).

In the majority of HIV-infected people, CD4 counts are declined due to poor medication adherence, substance use, and poor nutrition (Montarroyos et al., 2014). Accordingly, a team of counselors and nurses provided a healthy dietary schedule and health education regarding treatment, prognosis, and medication adherence with regard to HIV/AIDS.

Some biological, mental, and social issues are common to all HIV-infected cases, such as perceived stigma, fear of death, and negative attitudes of societal members. Those common issues of PLHA were resolved using universalization techniques of social casework (Mathew, 1992), according to which PLHA restructured their cognition when they see many other people affected by HIV/AIDS. The care and support group of the Jharkhand network of positive people facilitated through support group meetings, peer support, and supportive therapies greatly benefited the PLHA. Earlier evidence documented the benefits of peer support (Kumakech et al., 2009) and supportive therapies (Maldonado et al., 1996) because they instilled a feeling of acceptance among PLHA.

However, the intensity and nature of psychosocial issues vary from case to case. Therefore, an individual tailored psychosocial intervention can improve the conditions of people who have HIV/AIDS, and a fit-for-all system cannot resolve all the issues related to HIV/AIDS.

2.4.1 Limitation and Future Direction

Selection of six participants in the study was not randomized, there was no control group, and the study was not blinded. It was tough to find a control group among PLHA, where there will be no therapeutic intervention. Administering of scales and psychotherapy was carried out by the first researcher. Another team executed the rest of the intervention package. This comprehensive intervention package can be applied in randomized control trials among PLHA.

2.4.2 Implication and Contribution

Augmentation of psychosocial management of HIV/AIDS with antiretroviral therapy will improve medication adherence, CD4+ T cell, and the efficacy of ART.

References

Attrill, R., Kinniburgh, J., & Power, L. (2001). *Social exclusion and HIV: A report*. London, UK: Terrence Higgins Trust.

Beck, A. T., Steer, R. A., & Carbin, M. G. (1988). Psychometric properties of the Beck Depression Inventory: Twenty-five years of evaluation. *Clinical Psychology Review, 8*(1), 77–100. https://doi.org/10.1016/0272-7358(88)90050-5

Bienvenu, M. J. (1970). Measurement of marital communication. *The Family Coordinator, 19*(1), 26. https://doi.org/10.2307/582142

Bouman, T. K., & Visser, S. (1998). Cognitive and behavioural treatment of hypochondriasis. *Psychotherapy and Psychosomatics, 67*(4–5), 214–221.

Breet, E., Kagee, A., & Seedat, S. (2014). HIV-related stigma and symptoms of post-traumatic stress disorder and depression in HIV-infected individuals: does social support play a mediating or moderating role? *AIDS Care, 26*(8), 947–951. https://doi.org/10.1080/09540121.2014.901486

Brown, P., Lyson, M., & Jenkins, T. (2011). From diagnosis to social diagnosis. *Social Science & Medicine, 73*(6), 939–943.

Catalan, J., Harding, R., Sibley, E., Clucas, C., Croome, N., & Sherr, L. (2011). HIV infection and mental health: Suicidal behaviour—Systematic review. *Psychology, Health & Medicine, 16*(5), 588–611. https://doi.org/10.1080/13548506.2011.582125

Chan, K., Stoové, M. A., & Reidpath, D. D. (2008). Stigma, social reciprocity and exclusion of HIV/AIDS patients with illicit drug histories: A study of Thai nurses' attitudes. *Harm Reduction Journal, 5*(1), 28. https://doi.org/10.1186/1477-7517-5-28

Collins, R. L., & Lapp, W. M. (1992). The Temptation and Restraint Inventory for measuring drinking restraint. *Addiction, 87*(4), 625–633. https://doi.org/10.1111/j.1360-0443.1992.tb01964.x

Cuijpers, P., Sijbrandij, M., Koole, S., Huibers, M., Berking, M., & Andersson, G. (2014). Psychological treatment of generalized anxiety disorder: A meta-analysis. *Clinical Psychology Review, 34*(2), 130–140. https://doi.org/10.1016/J.CPR.2014.01.002

Cuijpers, P., van Straten, A., Andersson, G., & van Oppen, P. (2008). Psychotherapy for depression in adults: A meta-analysis of comparative outcome studies. *Journal of Consulting and Clinical Psychology, 76*(6), 909–922. https://doi.org/10.1037/a0013075

Deering, K. N., Vickerman, P., Moses, S., Ramesh, B. M., Blanchard, J. F., & Boily, M. C. (2008). The impact of out-migrants and out-migration on the HIV/AIDS epidemic: A case study from south-west India. *AIDS, 22*(5), S165–S181.

Durvasula, R., & Miller, T. R. (2014). Substance abuse treatment in persons with HIV/AIDS: Challenges in managing triple diagnosis. *Behavioral Medicine, 40*(2), 43–52. https://doi.org/10.1080/08964289.2013.866540

Engel, G. (1977). The need for a new medical model: A challenge for biomedicine. *Science, 196*(4286), 129–136. https://doi.org/10.1126/science.139.3550.81

Fink, P., Ewald, H., Jensen, J., Sørensen, L., Engberg, M., Holm, M., & Munk-Jørgensen, P. (1999). Screening for somatization and hypochondriasis in primary care and neurological in-patients: a seven-item scale for hypochondriasis and somatization. *Journal of Psychosomatic Research, 46*(3), 261–273. Retrieved from http://www.ncbi.nlm.nih.gov/pubmed/10193917

Goforth, H. W., Lowery, J., Cutson, T. M., Kenedi, C., & Cohen, M. A. (2009). Impact of bereavement on progression of AIDS and HIV infection: A review. *Psychosomatics, 50*(5), 433–439. https://doi.org/10.1016/S0033-3182(09)70834-5
Halli, S. S., Blanchard, J., Satihal, D. G., & Moses, S. (2007). Migration and HIV transmission in rural South India: An ethnographic study. *Cultural Health Sex, 9*, 85–94.
Hamilton, M. (1959). The assessment of anxiety states by rating. *British Journal of Medical Psychology, 32*(1), 50–55. https://doi.org/10.1111/j.2044-8341.1959.tb00467.x
Hintze, J., Templer, D. I., Cappelletty, G. G., & Frederick, W. (1993). Death depression and death anxiety in HIV-infected males. *Death Studies, 17*(4), 333–341. https://doi.org/10.1080/07481189308252629.
Hosseinian, S., Tabatabaei, S., & Ravaei, F. (2011). Effectiveness of supportive psychotherapy on quality of life of women with HIV in Kermanshah. *Procedia - Social and Behavioral Sciences, 30*, 1902–1907. https://doi.org/10.1016/j.sbspro.2011.10.370
Jayasvasti, D., Hiransuthikul, N., Pityaratstian, N., Jayasvasti, I., Lohsoonthorn, V., Kanchanatawan, B., & Triruangworawat, B. (2011). The effect of cognitive behavioral therapy and changes of depressive symptoms among Thai adult HIV-infected patients. *Article in World Journal of AIDS, 1*, 15–22. https://doi.org/10.4236/wja.2011.12003
Kumakech, E., Cantor-Graae, E., Maling, S., & Bajunirwe, F. (2009). Peer-group support intervention improves the psychosocial well-being of AIDS orphans: Cluster randomized trial. *Social Science & Medicine, 68*(6), 1038–1043. https://doi.org/10.1016/j.socscimed.2008.10.033
Kumar, R., Suar, D., & Singh, S. K. (2017). Regional differences, socio-demographics, and hidden population of HIV/AIDS in India. *AIDS Care, 29*(2), 204–208. https://doi.org/10.1080/09540121.2016.1211605
Kumar, R., Suar, D., Singh, S. K., & Bhattaharya, S. D. (2019). Why do AIDS-sufferers on antiretroviral therapy die early? Evidence from Jharkhand in India. In J. J. Kronenfeld (Ed.), *Research in the sociology of health care (Volume 38)*. Bingley, England: Emerald Group Publishing Limited.
Mackelprang, R. D., Bosire, R., Guthrie, B. L., Choi, R. Y., Liu, A., Gatuguta, A., … Farquhar, C. (2014). High rates of relationship dissolution among heterosexual HIV-serodiscordant couples in Kenya. *AIDS and Behavior, 18*(1), 189–193. https://doi.org/10.1007/s10461-013-0529-6
Maldonado, J., Gore-Felton, C., Durán, R., Diamond, S., Koopman, C. A. M., & Spiegel, D. (1996). *Supportive-expressive group therapy for people with HIV infection: A primer*. Stanford, CA: Psychosocial Treatment Laboratory Stanford University School of Medicine.
Mathew, G. (1992). *An introduction to social case work*. Mumbai, India: Tata Institute of Social Sciences.
Montarroyos, U. R., Miranda-Filho, D. B., César, C. C., Souza, W. V., Lacerda, H. R., Albuquerque, M. F. P. M., … Ximenes, R. A. D. A. (2014). Factors related to changes in CD4+ T-cell counts over time in patients living with HIV/AIDS: A multilevel analysis. *PLoS One, 9*(2), e84276. https://doi.org/10.1371/journal.pone.0084276
National-AIDS-Control-Organization. (2007). *Antiretroviral therapy guidelines for HIV-infected adults and adolescents including post-exposure prophylaxis*. New Delhi: Department of Health and Family welfare, Government of India. Retrieved from http://naco.gov.in/sites/default/files/1.Antiretroviral Therapy Guidelines for HIV-Infected Adults and Adolescents Including Post-exposure.pdf
Nebhinani, N., & Mattoo, S. K. (2013). Psychotic disorders with HIV infection: A review. *German Journal of Psychiatry, 16*(1), 43–48. Retrieved from http://www.gjpsy.uni-goettingen.de
Olley, B. O., Seedat, S., & Stein, D. J. (2006). Persistence of psychiatric disorders in a cohort of HIV/AIDS patients in South Africa: A 6-month follow-up study. *Journal of Psychosomatic Research, 61*(4), 479–484. https://doi.org/10.1016/j.jpsychores.2006.03.010
Primeau, M. M., Avellaneda, V., Musselman, D., St. Jean, G., & Illa, L. (2013). Treatment of depression in individuals living with HIV/AIDS. *Psychosomatics, 54*(4), 336–344. https://doi.org/10.1016/j.psym.2012.12.001
Richmond, M. E. (1917). *Social diagnosis*. New York, NY: Russell Sage Foundation.

Robertson, K., Bayon, C., Molina, J.-M., McNamara, P., Resch, C., Muñoz-Moreno, J. A., … van Wyk, J. (2014). Screening for neurocognitive impairment, depression, and anxiety in HIV-infected patients in Western Europe and Canada. *AIDS Care, 26*(12), 1555–1561. https://doi.org/10.1080/09540121.2014.936813

Saggurti, N., Mahapatra, B., Swain, S. N., Battala, M., Chawla, U., & Narang A. (2011). In N. UNDP and Population Council (Ed.), *Migration and HIV in India: Study of select districts*. New Delhi. Retrieved from http://www.in.undp.org/content/dam/india/docs/migration_and_hiv_in_india__study_of_select_districts.pdf

Sellman, J. D., Sullivan, P. F., Dore, G. M., Adamson, S. J., & MacEwan, I. (2001). A randomized controlled trial of motivational enhancement therapy (MET) for mild to moderate alcohol dependence. *Journal of Studies on Alcohol, 62*(3), 389–396. https://doi.org/10.15288/jsa.2001.62.389

Shamu, S., Zarowsky, C., Shefer, T., Temmerman, M., & Abrahams, N. (2014). Intimate partner violence after disclosure of HIV test results among pregnant women in Harare, Zimbabwe. *PLoS One, 9*(10), 1–8.

Spies, G., Asmal, L., & Seedat, S. (2013). Cognitive-behavioural interventions for mood and anxiety disorders in HIV: A systematic review. *Journal of Affective Disorders, 150*(2), 171–180. https://doi.org/10.1016/j.jad.2013.04.018

Thomson, A., & Page, L. (2007). Psychotherapies for hypochondriasis. In A. Thomson (Ed.), *Cochrane database of systematic reviews*. Chichester, UK: Wiley. https://doi.org/10.1002/14651858.CD006520.pub2

UNDP PFI, N. (2011). In U. N. D. Project (Ed.), *Migration and HIV in India: Study of select districts*. New Delhi: National AIDS Control Organization. Retrieved from http://www.in.undp.org/content/dam/india/docs/migration_and_hiv_in_india__study_of_select_districts.pdf

Wicks, A. L. (1996). Ethnicity and psychotherapy with HIV-infected patients: Three case studies. *Journal of Social Distress and the Homeless, 5*(1), 67–82.

World-Health-Organization. (1992). *The ICD-10 classification of mental and behavioral disorders: Clinical descriptions and diagnostic guidelines (ICD-10)*. Geneva. Retrieved from https://www.worldcat.org/title/icd-10-classification-of-mental-and-behavioural-disorders-clinical-descriptions-and-diagnostic-guidelines/oclc/28294867?page=citation.

World-Health-Organization. (2004). *Patient monitoring guidelines for HIV care and antiretroviral therapy (ART)*. Geneva. Retrieved from https://www.who.int/hiv/pub/guidelines/patientmonitoring.pdf

Chapter 3
Blended Learning and Its Impact on Cognition and Emotion

Preethi Nanjundan and Zahida Mariyam

Contents

3.1 Introduction

The learning landscape is dynamic and fast changing; the lesson planning and delivery have been more learner centric and are designed to suit individual's learning, skills, goals, and attitude. A decade ago, there was a tug of war for the online and the traditional learning. The tussle continues to find common ground in terms of treatment, elements, and components. However, after witnessing the benefits of technology-integrated lessons, instructors and instructional designers have called a truce and are working together to create a perfect blend of the best of various learning modalities. The intent is to create an **experiential** and **engaging** learning environment.

AI has been the driving force in bringing about cutting-edge and highly effective transitions in the educational sector, benefiting both students and the institutions.

P. Nanjundan
Department of Computer Science, Christ (Deemed to be University), Lavasa, Pune, India
e-mail: preethi.n@christuniversity.in

Z. Mariyam (✉)
Cambridge Certified Online English Trainer, ESL Trainer, Bangalore, Karnataka, India

S. N. Mohanty (ed.), *Emotion and Information Processing*,
https://doi.org/10.1007/978-3-030-48849-9_3

3.1.1 Blended Learning

Blended learning refers to a combination of offline (traditional learning methodologies, face-to-face learning) and online learning strategies (online quizzes, discussion boards, and other resources) in perfect synchronization with each other. Also known as mixed or hybrid learning, this learning approach is versatile and can be put to use in various forms. Supportive technology, educational software, and new learning tools have aided educators across the world, to enjoy the many advantages of blending in-class and online learning strategies, giving rise to this interesting model known as blended learning.

The prime focus of the blended learning model is to give learners more freedom in the way that they learn and engage in their education.

3.1.2 Learning Platforms

There is an ever-growing list of learning platforms that can be combined to achieve flexibility in designing tailor-made lessons to effectively fulfill learning objectives (Table 3.1).

3.1.3 Blended Learning types

There are six methods of **blended learning** as discussed below:

(a) **Online Model**
In this model, an instructor delivers lessons via an online platform using various eLearning resources complemented by periodic face-to-face meets. UK's first virtual classroom, 'Oxford Hub for International Virtual Education (HIVE)', is the first of its kind and second installation in the world. This platform allows university lecturers to engage with 84 students joining from across

Table 3.1 Some of the components that can be blended into an effective learning curriculum

Classroom training	Virtual classrooms (interactive web-based training)
Instructor lead training	Self-paced training
Print-based workbooks	Interactive interface
Seminars and Workshops	On-demand learning and collaboration (GoToWebinar, ClickMeeting, Skype)
Distance Learning	Massive Open Online Course (MOOC)
On-the-job-training	Learning groups and peer share forums
Podcast	Gamification of lessons (Kahoot is a well-known game-based learning technology)

the globe as if they were physically in the same class. The combination of robotics, facial recognition, and real-time video communication tools have resulted in creating an immersive learning and teaching experience at HIVE.

(b) **Flex Model**

In this model, learners are given customized, flexible schedules on an individual basis for online work, along with other course materials. The flex model involves an on-site instructor for teaching the online content while also providing face-to-face support for individual instruction, small-group instruction, as well as group projects.

This model can be effectively blended in a classroom. Students can be grouped and given tasks on laptops, and a teacher is present not to teach but to facilitate the learner. They clear any doubts the students might have and also discuss the work submitted by the groups.

(c) **Rotation Model**

In this model, we have two options:

Station Rotational Model. Students go from one learning activity to another, either in a structured learning session managed by the instructor or online in a self-directed fashion. Between different learning stations, one of which is an online station and the others include activities such as small-group or full-class instruction, group projects, individual tutoring, and pencil-and-paper assignments.

Lab Rotational Model. Students rotate at fixed schedules or at the instructor's discretion between classroom and a computer lab. Here, most of the learning happens online, and the classroom is generally reserved for other learning activities.

(d) **Self-blend Model**

This model is highly popular in high schools. The students, who are interested in learning beyond what is taught at school, supplement their learning through online courses offered remotely; this type of learning includes various resources such as videos, webinars, and blog content.

(e) **Online Lab**

This is another popular blended learning model which is entirely digital, with minimal or no instructor interaction. It usually takes place before, during, or after training. In this model, learners can access content on their mobile phones (m-Learning), laptops, or other smart devices for better engagement and solid learning. This model of blended learning is ideal for students who need more flexibility and independence in their daily schedules. Students can usually chat with their instructor online if they have any questions. Futurelearn.com is a fitting example for this model.

(f) **Face-to-Face Learning Model**

Of all the blended learning models, face-to-face driver is the closest to a typical school structure. In this model, lessons are delivered using online media by an on-site instructor. It is also sometimes referred to as Instructor-Led Training (ILT) or Classroom Training. This approach helps students who are struggling or working above their grade level to progress at their own pace using technology in the classroom.

3.1.4 Advantages of Blended Learning

In today's fast-paced world, the trend is to use different forms of training with a spectrum of audience. Learning is not limited within the four walls of schools, universities, or professional institutes. It is also a vital HR function of any organization, which adopts a growth mindset. However, with the increasing use of technology in learning, striking a balance between the right tool and correct instructional approach has become a major challenge and calls for the expertise of the professionals. Clarity of objectives, training options, and analyzing the correct solutions to any learning needs are essential to craft cost-effective, goal-oriented, and skill-enhancing courses.

- Blended learning offers the best of multiple modalities, giving the flexibility to create personalized learning experiences.
- Blended learning can offer a 24/7 online access to training resources, thereby removing the time and location limitations.
- Tracking learning and performance can become convenient, especially when using online assessments.
- One of the main advantages of blended learning is the cost effectiveness, improving the feasibility and return on investment (ROI) of training. Blended learning can be instrumental in slicing into a chunk of expenses associated with traveling, logistics, and trainer fees.
- Standardization is a core issue of any learning, and instructional designers make it one of their prime objectives to standardize all courses across various platforms. Blended learning components make it possible to achieve this through its online and virtual learning segments.

3.1.5 Blended Learning in the Workplace

Blended learning at the workplace is a solution-based approach which is both time- and cost-effective. It has become an integral part of organizational learning program and is adopted to empower employees by improving their core skills and aligning individual competencies with team to fulfil business objectives. It is a widely used and valuable technique for learning in the workplace for a number of reasons. Given below are a few ways it can help employees in a corporate setup.

- Enhanced employee engagement as employees have control over their training with the ability to use smart devices or laptops when and where it is convenient for them.
- Personalized lesson plans which can be tailor-made to suit individual's training needs. Placement tests can be used to decide modules where more work is required and cut down on those which the learner has gained mastery over.

- More effective training feedback and opportunities to further improve the learning program as it allows the collection of valuable data through tests, online quizzes, etc.
- Cost-effective and time-efficient, blended learning enables institutions to provide personalized training experience to learners at a fraction of the cost and time. L&D teams can save time in planning out curricula by adapting training programs designed by MOOCs and online course providers with proven results.
- Increased satisfaction as blended learning accommodates different learners and learning styles when compared with face-to-face learning.
- Better retention of information, owing to the flexibility of accessing the lessons anytime and anywhere, learner can revisit the material after class as and when required, thus cementing the information.

3.1.6 Reasons Why Blended Learning Is Effective

There are various reasons that make **blended learning** effective and popular. Some of these are

- The approach caters to different learning styles and helps in improving the performance.
- Offers increased flexibility as it enables anytime anywhere learning.
- It is an inclusive model that has visual, written, as well as auditory components to cater to every type of learner.
- It allows the participants to learn at their own pace and their own ability level.
- The model eliminates the need to physically attend classes, thus benefiting students who cannot adhere to a fixed schedule.
- Having blend of different learning components assures learners' engagement resulting in effective outcomes.

3.1.7 Power of Audio-Visual Medium

Dale's Cone, it explains a lot about the impression that a visual medium has on learning. Dale explained the various modalities of assimilation of learning and defined their impact (Fig. 3.1).

During 1960s, Edgar Dale hypothesized that learners retain more information by what they "do" as opposed to what is "heard," "read," or "observed." His research was the stepping stone in the development of the Cone of Experience. Today, the term "learning by doing" has been termed as "experiential learning" or "active learning." According to Dale's research, the least effective method at the top involves learning from information presented through verbal symbols, i.e., listening to lectures. The most effective methods at the bottom involves direct purposeful

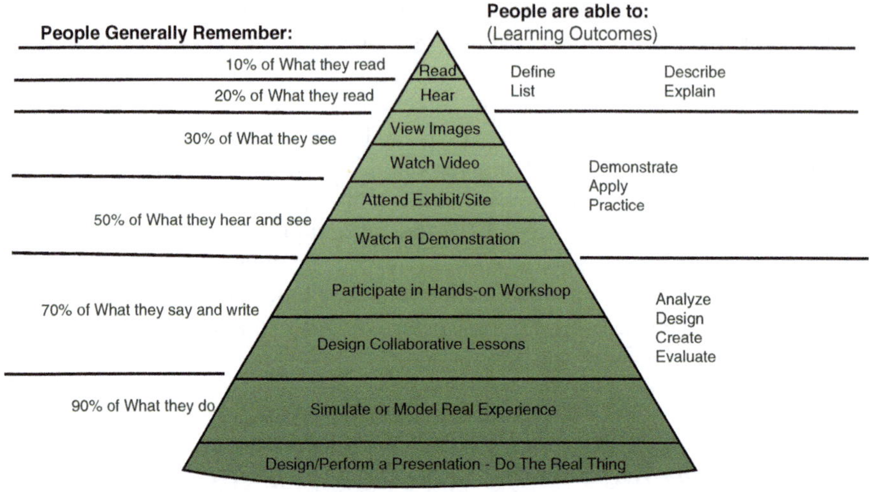

Fig. 3.1 Source: Adapted from E. Dale, Audiovisual Methods in Teaching, 1969a, b, NY: Dryden Press

learning experience, such as hands-on or field experience. Direct purposeful experience represents reality or simulation of everyday life (Anderson, 2003). The study concludes that "active learning" techniques result up to 90% retention. Learners learn best when they are actively engaged in the learning.

Dale's Cone theory stresses on the fact that active learning is experiential, mindful, and active learning (Mohanty & Suar, 2013). Active learning requires students to participate in class, as opposed to passively listening. Modern classroom teaching integrates activity-based learning thanks to technology integration and smart platforms which allow educators to create quizzes, cartoons, and live worksheets making learning fun and productive.

3.1.8 AI a Game Changer

The impact of Artificial Intelligence in education is being felt by the widespread and rapid change in the traditional methods of teaching and assessments. The academic world is becoming easy to use and personalized thanks to the numerous applications of AI for education. This has brought about a significant change in the way people learn. Educational materials are now digitalized and are accessible across the globe through smart devises and computers. Today, virtual classrooms are taking over the traditional ones; as long as learners have computers and internet connection, they can attend class from any corner of the world. AI is also aiding the automation of administrative tasks, allowing institutions to minimize the time required to complete difficult and repetitive tasks and allowing educators to spend more time with the students.

The transformations brought by AI in Education are:

1. **Simplifying Administrative Tasks**

   AI can automate the expedition of administrative duties for teachers and academic institutions. Educators were initially spending a lot of time grading exams, assessing homework, and providing valuable feedback to their students. But technology has made it possible to automate the grading tasks where multiple tests are involved. This has resulted in professors having more time with their students rather than spending long hours correcting their test papers. Software providers are coming up with smarter ways of grading written answers and normal essays. AI is also proving highly resourceful for school admissions board. Artificial Intelligence is streamlining the entire admin process by automating the classification and processing of paperwork.
2. **Smart Content**

   AI has been a catalyst in a new technique of education called "Blended Learning"; education goes hand in hand with technology-aided lessons. These new techniques could be all that is required to ensure that learners attain their ultimate academic success. Smart content is a very hot subject matter today. This technology has already reached a classroom setting and corporate training department. Smart content also includes virtual content like video conferencing and video lectures. Textbooks are taking a new turn with AI systems are using traditional syllabuses to create customized lessons for certain subjects. As a result, textbooks are being digitized, and new interactive learning interfaces are being created to help students of all academic grades and ages. An example of such mechanisms is the Kitaboo which uses AI to help you differentiate the design and delivery of your training content. Kitaboo is a cloud-based eLearning development platform that helps you create interactive multimedia content with DRM protection.
3. **Personalized Learning**

   Education is no longer just about taking notes and learning by the rote. Today, innovative educators in both academic and corporate learning and development are constantly improving learning through technology, as evidenced by the rapid adoption of technology-assisted teaching methods and blended learning models.

   The best part is that by the introduction of AI, teachers are not replaced, but they are in a position to perform much better by offering personalized recommendations to each pupil. AI customizes in-class assignments as well as final exams, ensuring that students get the best possible assistance.

   The power of blended learning methods lies in its ability to enrich the student experience. "Blended learning" has proven effective in reducing failure rates, improving learning, and boosting learner engagement. Blended learning combines the best aspects of face-to-face teaching and online instruction in ways that enable students to learn at their own pace, which is a boon when there are mixed ability learners in the group. For example, a student in a blended learning course who masters a concept earlier than his peers can move on without having to wait, and similarly, a student who needs more time is not forced to move forward before fully grasping the subject. It is proving to be a customizable learning model that can be extended to work for diverse populations of students. Thanks to AI, smart tutoring systems, like Write& Improve platform developed by Cambridge University, which offers quick feedback and works directly with students. Even though these methods are still in their inception stages, they will soon become fully-fledged digital teachers to assist students with any educational needs.

4. **Global Learning**

   Education has no limits, and AI has begun to eliminate boundaries. Technology brings drastic transitions by facilitating the learning of any course from anywhere across the globe and at any time. MOOCs and many online interactive learning portals have made possible to take up diploma or a degree program from any part of the world. AI-powered education equips students with fundamental IT skills. With more inventions, there will be a wider range of courses available online, and with the help of AI, students will be learning from wherever they are.

5. **Making Lessons More Memorable: Cognitive Load Theory (CLT)**

Cognitive Load Theory (CLT) is an instructional design theory that uses the science to explain how the human brain processes information and applies the same to the design of learning materials to ensure they are easy to comprehend and remember.

In the year 1968, Richard Atkinson and Richard Shiffrin proposed the model of information processing, on which CLT is based. The model explains how information is exchanged between our three repositories: our sensory memory (short-term memory), our working memory, and our long-term memory. Armed with the findings from this research, John Sweller in 1988 developed CLT after analyzing the model presented by the two Richards. "Cognitive load" is the amount of information a human's working memory can process at one time. Sweller stated that because working memory is limited, it should not be overloaded unnecessarily during the learning process.

CLT states that working memory can be extended. Human brain processes visual and auditory information separately. For example, an image with additional information provided through audio narration delivers less cognitive load than two visual items. An existing schema is treated as a single item by your working memory, whereas a well-practiced schema that has become automated hardly counts as an item at all.

The lessons developed using AI-aided platforms which use multimedia's help to create powerful, interactive, and well-balanced lessons result in learning messages which are understood and remembered (Fig. 3.2).

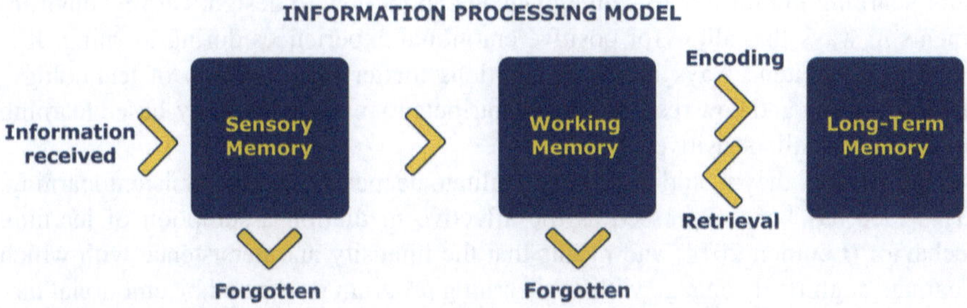

Fig. 3.2 The Human Brain Process Information (Atkinson & Shiffrin, 1968)

3.2 Emotional States and Learning Outcomes

Educational research has always recognized the importance of taking emotions into account when analyzing any learning process. Efforts are made to make all interactions with computers and other electronic media fundamentally social and natural, similar to real life. But surprisingly, little has been done in terms of research to study emotions in a blended learning context. Emotional states govern thought and behavior and are responsible for all brain–body actions. A learner's emotional state is the number one influencer on his or her learning behavior. Smart managing of emotional states will result in the learner having an easier time, grasping and retaining the lesson.

The brain seeks stimulation of novelty and the safety of familiar things. Novelty piques interest and generates enthusiasm, thus acting as a positive stimulation to seek out information and start the learning process. On the other hand, brain also finds safety in familiar things. The emotional receptivity to learning can be highly improved if allowances are made for both the responses during the course of teaching. Safety is found in repetition of activities until they become familiar. These repetitive tasks can be termed as "Rituals." The best rituals are those which are predictable, create unity, are easy for everyone, and effortlessly put learners in a productive state. Movement, humor, and music have the capacity to put the learner in a positive brain state. Learning can be accelerated by using these mood enhancers (Mohanty, Pratihar, & Suar, 2015).

3.2.1 Is There Any Cognitive and Emotional Link Connecting These Two to Online Learning?

In most technology-based learning environments, the learning goals are cognitive. For successfully reaching these goals, however, the emotions a learner experiences during learning play an essential role. The challenge faced by course designers in an e-learning platform is to gain knowledge about how to design learning environments in ways that allow for positive emotional experiences during learning. It is vital to understand ways in which emotions matter in the design of technology-based learning and how research may contribute to make technology-based learning more emotionally sensitive.

In this way, they would be able to facilitate deeper and more persistent learning. This idea has been expressed as the affective mediation assumption of learning behavior (Leutner, 2014) and means that the intensity and persistence with which learners cognitively engage with any learning program is affected by emotional factors. Play of Emotions in Game-Based Learning: engaging the inner child gamification has proved to be effective for learning because of these four elements which do not elicit a cognitive response but also engage emotions.

3.2.2 Retrieval

The first is the practice of retrieval which requires the learner to answer quiz questions about a topic. The benefits of retrieval process are known for at least 100 years, and the results have been the same with many diverse groups. The gamification platforms do not just ask the learner to recall content and act on that recall but also emotionally stimulate the learner by use of constant praise each time the learner responds correctly, making a grand full-out. Digital celebration each time the learner completes a full lesson and keeping full track of the learner's progress though point schemes. Few apps also challenge the learner by betting points against them that they would not keep up the same learning streak up for another week. Therefore, it successfully challenges the learner to prove their mettle and makes sure the excitement toward learning is constant.

3.2.3 Spaced Retrieval

The second practice involves with course content and quiz spaced over time, delivered daily or weekly most often through a mobile device. This is also one of the proven techniques by educational psychology research. The greater the amount of spacing between retrieval of information (24 h is optimal), the higher the possibility of retention (Carpernter & DeLosh, 2005). Spaced retrieval is made fun and convenient by end of the week quiz, and learner is also provided a summary of the lessons learnt which can be used to revise before attempting the quiz. Once again welcoming the challenge of keeping one's score up can be a positive incitement to revise the lessons, thus gaining deeper understanding of the learned material.

3.2.4 Heightened Attention Level

Neuroscientific research on memory has revealed that the process of storing new memories can be significantly enhanced by increasing physiological arousal (Cahill & McGaugh, 1998; Hamann, 2001). In layman's terms, engaging in a gamified learning situation stimulates learners by keeping their excitement levels high and improving their vigilance and focus for the subsequent learning material.

3.2.5 Challenge

Challenge has always proved to being a strong motivator in learning (Jones et al., 1994; Malone, 1981; Schlechty, 1997). The challenge is to engage a learner in a difficult but achievable task. Games are addictive, and they become more so when the learner is helped to set daily goals, reminder prompts, and rewards after each test is completed. It tackles the biggest challenge faced by the learners—motivation to continue learning.

References

Anderson, H. (2003). Dale's cone of experience. Retrieved from http://www.etsu.edu/uged/etsu1000/documents/Dales_Cone_of_Experience.pdf accessed on 2013-04-19.

Atkinson, R. C., & Shiffrin, R. M. (1968). Human memory: A proposed system and its control processes. In K. W. Spence & J. T. Spence (Eds.), *The psychology of learning and motivation* (Vol. 2, pp. 89–195). New York, NY: Academic Press.

Cahill, L., & McGaugh, J. L. (1998). Mechanisms of emotional arousal and lasting. *Trends Neuroscience, 21*(7), 294–299.

Carpenter, S. K., & DeLosh, E. L. (2005). Application of the testing and spacing effects to companies using AI in education to enhance the classroom. *Applied Cognitive Psychology, 19*(5), 619–636.

Dale, E. (1969a). *Audio-visual methods in learning* (3rd ed.). New York, NY: The Dryden Press.

Dale, E. (1969b). *Audio-visual methods in teaching* (3rd ed., p. 108). New York, NY: Holt, Rinehart & Winston. December 3, 2018, Updated: February 24, 2020, Written by Alyssa Schroer, Builtin.com.

Hamann, S. (2001). Cognitive and neural mechanisms of emotional memory. *Trends in Cognitive Sciences, 5*(9), 394–400.

Mohanty, S. N., & Suar, D. (2013). Influence of mood states, group discussion, and interpersonal comparison on change in decision-making and information processing. *International Journal of Decision Sciences, Risk and Management, 5*(2), 101–123.

Mohanty, S. N., Pratihar, D. K., & Suar, D. (2015). Influence of mood states on information processing during decision making using fuzzy reasoning tool and neuro-fuzzy system based on Mamdani approach. *International Journal of Fuzzy Computation and Modelling, 1*(3), 252–268.

Chapter 4
Theory and Implications of Information Processing

G. Sucharitha, Anjanna Matta, Kanagala Dwarakamai, and Bodepu Tannmayee

Contents

4.1 Introduction

The sciences involved gathering, manipulation, storage, retrieval, and classification of recorded information. It is the alteration (processing) of the information detectable in any way by an observer. As such, it is a mechanism that explains all that is occurring (changes) in the world, from dropping a rock (a shift in position) to printing a text file from a digital computing device. In the latter case, the form of presentation of that text file (from bytes to glyphs) is changed by an information processor (printer).

Information processing is the process of changing or converting information into a meaningful form. Information is a processed, organized, or classified data that is useful for the receiver. Information is the processed data that may be used "as is" or

The original version of this chapter was revised. The correction to this chapter is available at https://doi.org/10.1007/978-3-030-48849-9_14

G. Sucharitha (✉) · A. Matta · K. Dwarakamai · B. Tannmayee
Faculty of Science and Technology, ICFAI Foundation for Higher Education, Hyderabad, India
e-mail: sucharithasu@ifheindia.org; anjireddyiith@ifheindia.org

S. N. Mohanty (ed.), *Emotion and Information Processing*,
https://doi.org/10.1007/978-3-030-48849-9_4

may be put to use along with more data or information. The receiver of information takes actions and decisions based on the information received. The collected data must be processed to get meaning out of it, and this meaning is obtained in the form of information.

4.2 The Emergence of Information Processing

Information processing is a model for human thinking and learning, and it is a part of the resurgence of cognitive perspectives of learning. The cognitive perspective asserts that complex mental states affect human learning and behavior, that such mental states can be scientifically investigated. Computers, which process information, include the internal states that affect processing. Computers, therefore, provided a model for possible human mental states that provided researchers with clues and direction for understanding human thinking and learning as information processing. Overall, information-processing models helped in re-establish the mental processes that cannot be directly observed as a legitimate area of scientific research.

4.3 Theories of Information Processing

4.3.1 Levels of Processing

One of the first alternatives to the stage theory was developed by Craik and Lockhart (1972) and labeled the levels of processing model. Specifically, the levels of processing theory hold that memory is not three-staged which separates it immediately from the stage theory model. Craik and Lockhart argue that stimulus information is developed at several levels concurrently (not serially) depending on features, concentration, and meaningfulness. The innovative information does not have to enter in any specific order, and it does not have to pass through a prescribed channel. They further contend that the more deeply information is processed, the more that will be remembered (Kearsley, 2001b). This model was a precursor to the development of schema theory, discussed below. The two are consistent in that they agree that "the more connections to a single idea or concept, the more likely it is to be remembered" (Huitt, 2000).

4.3.2 Dual Coding Theory

As mentioned previously, another theory in the information processing debate is Paivio's work in dual coding (Clark & Paivio, 1991). This theory gives equal significance to both verbal and nonverbal processing and suggests that there are two separate systems for processing these types of information. Images—mental images—are processed by one system, and logo-gens—verbal entities, chunks, or propositions—are processed by a different system. According to Kearsley (2001a), Paivio believes that human cognition is unique in that it has become specialized for dealing simultaneously with language and with nonverbal objects and actions. Moreover, the language system is peculiar in that it deals directly with linguistic input and output (in the form of speech or writing), while at the same time serving a symbolic function concerning nonverbal objects, events, and behaviors. Any representational theory must accommodate this dual functionality. Further, Paivio suggests that there are three separate types of processing and communication between these two subsystems: representational, referential, and associative. Representational processing is the direct activation of one system or the other; referential is the activation of one sub-system by the other; and, associative is activation within the same subsystem without the interaction of the other.

4.3.3 Schema Theory, Parallel Distributed Processing, and Connectionist Models

Rumelhart (1980), working in conjunction with others, developed the schema theory of information processing and memory. He suggested that a schema is a data structure for representing generic concepts stored in memory.

There are five key components to this view of memory and processing in relation to schema.

1. It is an organized structure that exists in memory and is the sum of all gained knowledge.
2. It exists at a higher level, or abstraction, than immediate experience.
3. Its concepts are linked by propositions (verbal constructs).
4. It is dynamic.
5. It provides a context or structure for new information (Winn & Snyder, 2001).

This model is sometimes called the connectionist model or theory. Huitt (2000) explains that "This model emphasizes the fact that information is stored in multiple locations throughout the brain in the form of networks of connections." This model is explicitly different from previous ones in that it is not founded on the belief in a serial processing description. Rather, the connections between information are key, not the order in which connections are made. Rumelhart later worked with McClelland and the Parallel Distributed Processing Research Group (McClelland & Rumelhart, 1981, 1986; Rumelhart & McClelland, 1986) to expand his initial work and connectionist theories. In this enhanced model, it is still proposed that the units of memory are connections rather than any concrete representation of previous information. The latter model goes further, however, saying that the activation of the connections is the knowledge unit. According to Driscoll (2001), there are many advantages to this model. She says that it accounts for the incremental nature of learning is dynamic, incorporates goals of learning, and has the potential to explain cognitive development.

4.4 Types of Information Processing

Information can be processed in two ways.

4.4.1 Natural Information Processing System

Specific mechanisms for processing the information can be created in nature. Like all structures for the processing of information, their purpose is to coordinate information concerning the operation of organizations implemented by a program. Natural systems for processing information control the behaviors of natural beings, such as living organisms. There are many options to define the underlying logic of natural information processing processes, but we will concentrate on five basic principles in this chapter (see Table 4.1) to show how they relate to both human cognition and natural selection evolution. Some research evidence suggested local and global information processing may help on decision making under uncertainty (Mohanty & Suar, 2013a, b).

4.5 Technical Information Processing System

Since the introduction of the term "cybernetics" by Norbert Wiener (1948), computer science and ICT (also known as Information Processing Technology, IPT in this chapter) developed in an unpredictable and unimaginable manner, a phenomenon also known as the information revolution (Porter & Read, 1998) and recently also known as the "second information revolution" underlining the remarkable

Table 4.1 Principles of general information processing system

Principles	Cognitive study	Evolutionary study	Function
Information storage principle	Long-term memory	Genome	Accumulate information for imprecise phases
Borrowing and reorganizing principle	Transmit information to long-term memory	Transmit information to the genome	Permit the rapid building of an information store
Arbitrariness as genesis principle	Create new ideas	Generate new genetic codes	Generate new information
Slender limits of transform principles	Operational memory	Epigenetic system handling ecological information	Input ecological information to the information store
Environmental organizing and linking principle	Long-term operational memory	Epigenetic system handling genetic information	Use information stored in the information store

growth of online communication. The changes in communication processing situations that have occurred over the past decade are often referred to as Web 2.0 (O'Reilly, 2005), a buzzword that reflects technological and relational innovations. Finally, a consumer can communicate with multiple devices, for example, to exchange emails, apart from personal computers (PCs), one can use a mobile phone, a PAD (personal digital assistant), or a smartphone (e.g., an iPhone) that contains all functionalities, access to websites and a TV set may also be acceptable, etc.

From the online communication scenario, two major themes emerge:

1. Integration of communication tools and functionalities into the same interface for the website
2. App fusion to reach these networks, usually social network websites

4.5.1 Stages of Information Processing Cycle

Information processing is a sequence of events consisting of input, processing, storage, and output. To understand more about what is information processing cycle, it is a good idea to study about data processing cycle also. These events are similar to the case of the data processing cycle. For a computer to perform useful work, it has to receive instructions and data from the outside world. The computer receives data and instructions during the INPUT stage of the information processing cycle. Useful information results when appropriate instructions are applied to data. Applying instructions to data takes place during the PROCESSING stage of the information processing cycle. To avoid having to re-enter data and instructions or reprocess information, computers can save information. Saving information on a computer occurs during the STORAGE phase of the information processing cycle. Saving information on a computer occurs during the STORAGE phase of the information processing cycle. This is followed by the result in the OUTPUT stage. Computer Processing Cycle is a similar process with similar steps by which data is fed to a computer.

Input

1. Entering data into the computer.
2. Feeding the collected raw data in the cycle for processing. This is the raw data that is supplied for processing and obtaining information.
3. Input can be done by utilizing various devices such as keyboards, mice, flatbed scanners, barcode readers, joysticks, digital data tablets (for graphics drawing), and electronic cash registers.

Processing

1. Performing operations on the data.
2. Once the input is provided the raw data is processed by a suitable or selected processing method. This is the most crucial step as it allows for the processed data in the form of output which will be used further.

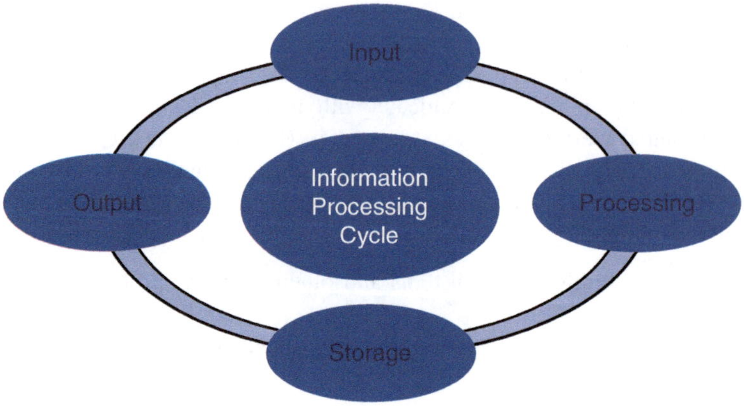

Fig. 4.1 Information processing cycle

3. Processing is usually done by CPU (central processing unit) in a computer. CPU is the crucial component for getting the operations done.

Storage

1. Saving data in a soft/physical form.
2. This is the outcome, and the raw data provided in the first stage is now "processed," and the data is useful and provides information and no longer called data.
3. Storage can be done on an external hard disk, inbuilt hard disk, pen drives, micro SD cards, compact disks, or even in registers.

Output

1. Results obtained, i.e., information.
2. This is the outcome, and the raw data provided in the first stage is now "processed," and the data is useful and provides information and no longer called data. This might be further used for data visualization.
3. This can be used as it is or used for further processing along with more data (Fig. 4.1).

4.6 Memory

One of the primary areas of cognition studied by researches is memory. There are many hypotheses and suggestions as to how this integration occurs, and many new theories have built upon established beliefs in this area. Currently, there is widespread consensus on several aspects of information processing; however, there are many dissentions about specifics on how the brain codes or manipulates information as it is stored in memory. Schacter and Tulving (as cited in Driscoll, 2001) state that, a memory system is defined in terms of its brain mechanisms, the kind of information it processes, and the principles of its operation. This suggests that memory is the combination of all mental experiences. In this light, memory is a built store that must be accessed in some way for effective recall or retrieval to occur. It is premised on the belief that memory is a multifaceted, if not multistaged, system of connections and representations that encompass a lifetime's accumulation of perceptions. Eliasmith (2001) defined that memory is the "general ability, or faculty, that enables us to interpret the perceptual world to help organize responses to changes that take place in the world." It is implied by this definition that there must be a tangible structure in which to incorporate new stimuli into memory. The form of this structure has been the source of much debate, and there seems to be no absolute agreement on what shape a memory structure takes, but there are many theories on what constitutes both the memory structure and the knowledge unit.

4.6.1 *The Stage Model*

Traditionally, the most widely used model of information processing is the stage theory model, based on the work of Atkinson and Shiffrin (1968). The key elements of this model are that it views learning and memory as discontinuous and multistaged. It is hypothesized that as new information is taken in, it is in some way manipulated before it is stored. The stage theory model recognizes three types or stages of memory: sensory memory, short-term or working memory, and long-term memory (Table 4.2 and Fig. 4.2).

4.6.2 *Sensory Memory*

Sensory memory represents the initial stage of stimuli perception. It is associated with the senses, and there seems to be a separate section for each type of sensual perception, each with its limitations and devices. Stimuli that are not sensed cannot be further processed and will never become part of the memory store. This is not to

Table 4.2 The three parameters of short- and long-term memories

Type	Short-term memory	Long-term memory
Capacity	7 ± 2 chunks of information	Unlimited
Duration	20–30 s	Permanent, long-term
Content	What you are currently thinking about (information from the sensory register and information from long-term memory)	Schemata

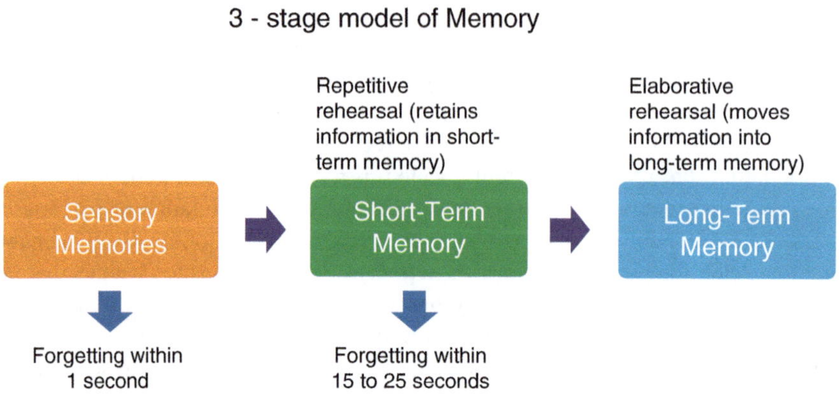

Fig. 4.2 Three-stage model of memory

say that only stimuli that are consciously perceived are stored; on the contrary, everyone takes in and perceives stimuli almost continuously. It is hypothesized, though, that perceptions that are not transferred into a higher stage will not be incorporated into memory that can be recalled. The transfer of new information quickly to the next stage of processing is of critical importance, and sensory memory acts as a portal for all information that is to become part of memory. This stage of memory is temporally limited which means that information stored here begins to decay rapidly if not transferred to the next stage. This occurs in as little as 0.5 s for visual stimuli and 3 s for auditory stimuli. There are many ways to ensure transfer and many methods for facilitating that transfer. To this end, attention and automaticity are the two major influences on sensory memory, and much work has been done to understand the impact of each on information processing. Attention is defined by Suthers (1996) as the "limitations in our perceptual processing and response generation: to attend to one this is to not attend to others." To attend to a stimulus is to focus on it while consciously attempting to ignore other stimuli, but it is not exclusive of these competing others. Treisman (as cited in Driscoll, 2001) "showed, however, that attention is not an all-or-nothing proposition and suggested that it serves to attenuate, or tune out, stimulation." Attention does facilitate the integration and transfer of the information being attended, but it is impacted by many factors including the meaningfulness of the new stimulus to the learner, the similarity between competing ideas or stimuli, the complexity of the new information, and the physical ability of the person to attend. Automaticity is almost the exact opposite of attention. Driscoll (2001) says that "When tasks are over learned or sources of information become habitual, to the extent that their attention requirements are minimal, automaticity has occurred." Automaticity allows attention to be redirected to other information or stimuli and allows for the ability of multitasking without distracting totally from the acquisition of new information. There are several suggested models of how new stimuli are recognized in sensory memory and how each deals with pattern recognition. The matching of new stimuli to the existing memory structures is a crucial factor in the acquisition of new knowledge. If new information is not brought into memory in a meaningful way, it will not be stored as memory. Therefore, the understanding of the patterns by which this information is represented is critical to the proper introduction of new information. Driscoll (2001) says that pattern recognition is "the process whereby environmental stimuli are recognized as exemplars of concepts and principles already in memory." She discusses three models of pattern recognition: template matching, the prototype model, and feature analysis. The template matching model holds that there are exact representations of previous stimuli trapped in the mind. Pattern recognition, then, occurs by matching input with a specific, perfect specimen stored in memory. This model seems to fall short because of the vast numbers of templates that would have to exist in memory for any one type of entity and because it does not account for imperfect stimuli or imperfect templates. The second pattern recognition model is the prototype. This model suggests that the stored unit is a generalized or abstracted form of the knowledge unit, and pattern recognition is based on a comparison of the input to the prototype. If a close match is established, new information can be accepted as the existing class. These two models are very similar in that they each attempt to match incoming information with a whole picture stored in memory. This holistic comparison differentiates them from the third model, feature analysis. In this system, incoming information is judged based on characteristics rather than a whole idea. Individual characteristics are picked out and then grouped to label the new stimulus as "X." The major difference is that these two models seem to work in opposite directions.

4.6.3 Short-Term or Working Memory

The second stage of information processing is the working or short-term memory. This stage is often viewed as active or conscious memory because it is the part of memory that is being actively processed while new information is being taken in. Short-term memory has a very limited capacity, and unrehearsed information will begin to be lost from it within 15–30 s if other action is not taken. Two main ways are effective in processing information while it is in short-term memory. Rote or maintenance rehearsal is the first but less desirable of these methods. This type of rehearsal is intended only to keep the information until it can be processed further. It consists mainly of some sort of repetition of the new information, and if it is not processed further, it will be lost. Studies on the limitations of working memory have revealed a specific number of units that the mind can process at any given time, and it is now generally accepted that 5 + 2 is the maximum number of stimuli that can be processed at once. There are several types of activities that one can perform to encode new information, but the importance of encoding cannot be overstated. Maintenance rehearsal schemes can be employed to keep information in short-term memory, but more complex elaboration is necessary to make the transfer to long-term memory. It is necessary for new information to somehow be incorporated into the memory structure in order for it to be retained. There are many suggested models for encoding, but there are three ways in which retention occurs. A stimulus can be an almost exact match with existing structures in which case it would be simply added to the mental representation and no change would be made to the structure except its addition. If the new stimulus does not exactly match the existing structure, the structure itself would be adapted to allow for additional characteristics or definitions in which case there would be a fundamental change to the existing structure, which would broaden the defining structures. Finally, if the new stimulus were vastly different from any existing structure, a new one would be created in memory. This new structure could in some way be linked to relevant structures, but it would stand alone as a new unit. At any rate, the incoming information must be acted on and through existing structures and incorporated into those systems in some way for acquisition to occur. The processing of this new stimulus takes place in short-term memory, and the body with which the information is worked is the long-term memory. The implications of this research are clear. If learning—relatively permanently change—is to take place, new information must be transferred into long-term memory. Therefore, repetition and maintenance rehearsal are not sufficient to produce a lasting effect. This has great relevance to instruction and teaching, for if the aim of education is learning, information must be presented in such a way that it can be incorporated into the memory structure. Long-term memory: As discussed with short-term memory, long-term memory houses all previous perceptions, knowledge, and information learned by an individual, but it is not a static file system that is used only for information retrieval. Abbot (2002) suggests that long-term memory "is that more permanent store in which information can reside in a dormant state—out of mind and unused—until you fetch it back into consciousness" (p. 1). In order to incorporate new information, long-term memory must be in communication with short-term memory and must be dynamic. There are several categories of long-term memory, and there are many suggestions as to how memory units are represented in the mind. While it seems that it might be sufficient to understand simply that some individual units and structures exist in long-term memory, the specific way or ways that information is stored offer extremely important information. If the knowledge unit is pictorial rather than verbal, for example, it would seem to make sense that images would be more easily and readily stored in memory. If the reverse were true, the information should be presented in verbal constructs. This oversimplifies the problem, but it is this question that is at the core of the controversy over memory storage structures. There are two divisions at issue in the discussion of long-term memory: the types of long-term memory and the type of knowledge unit stored in long-term memory. Organizations of long-term memory: Today, cognitive psychologists believe that there are at least different types of information stored in long-term memory. Each of the memory structures is distinct and serves a different operational function. However, it is evident that some type of very specialized categorization system exists within the human mind. One of the first to make this idea explicit was Bruner (as cited in Anderson, 1998). "Based upon the idea of categorization, Bruner's theory states 'To perceive is to categorize, to conceptualize is to categorize, to learn is to form categories, to make decisions is to categorize'."

4.7 Computer–Mind Analogy (Fig. 4.3)

The development of computer in the 1950s and 1960s had an important influence on psychology and was, in part, responsible for the cognitive approach becoming the dominant approach in modern psychology (taking over from behaviorism). The computer gave cognitive psychologists a metaphor, or analogy, to which they could compare human mental processing. The use of the computer as a tool for thinking how the human mind handles information is known as the computer analogy. Essentially, a computer codes (i.e., changes) information, stores information, uses information, and produces an output (retrieves information). The idea of information processing was adopted by cognitive psychologists as a model of how human thought works. For example, the eye receives visual information and codes information into electric neural activity which is fed back to the brain where it is "stored" and "coded." This information can be used by other parts of the brain relating to mental activities such as memory, perception, and attention. The output (i.e., behavior) might be, for example, to read what you can see on a printed page. Hence, the information processing approach characterizes thinking as the environment providing input of data, which is then transformed by our senses. The information can be stored, retrieved, and transformed using "mental programs," with the results being behavioral responses. Cognitive psychology has influenced and integrated with many other approaches and areas of study to produce, for example, social learning theory, cognitive neuropsychology, and artificial intelligence.

Fig. 4.3 The computer–mind analogy

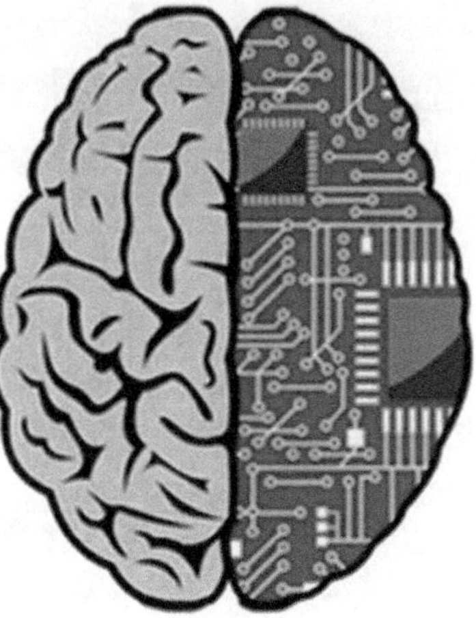

4.7.1 How Does Psychology Relate to Information Processing?

During the 1960s, American psychologists investigating and exploring the principles of cognitive theories ultimately developed a new approach called cognitive psychology or information processing. Cognitive psychology included a spectrum of processes like attention, perception, thinking, remembering, and problem-solving.

They fully gave up studying learning in isolation, and this resulted in studying human learning as a whole rather than its different components. The term cognition refers to the processes through which information coming from the senses is transformed (Fig. 4.4).

4.7.2 Metacognition

- Metacognition is our knowledge about attention, recognition, encoding, storage, and retrieval and how those operations might best be used to achieve a learning goal.

The Nature and Importance of Metacognition
- It contains what we know about how person variables, task variables, and strategy variables affect learning.
- Thus, it determines the extent to which students can be strategic learners.

Age Trends in Metacognition
- Primary grade children have limited knowledge of:
- Their memory capability.

Fig. 4.4 Natural way of information processing

- Factors that affect reading comprehension and recall—the need to tailor learning tactics to task demands—when they have learned something well enough that they can pass a test.
- Metacognitive knowledge develops with age, experience, and instruction.

4.8 Implications and Applications of Information Processing

The thoughts described above outline a framework that can potentially work in an accepted environment. When presented in this sense, we conclude the framework and be applied to human cognition by way of precisely describing nature with the aid of natural determination and, most particularly for our purposes. If this definition accurately characterizes human cognition, some consequences drift in fields such as education, training, and data transport for the character of human cognition and implementations like thinking, decision-making, problem-resolving, and preparation. It can additionally be claimed that from the current theory, the very core of human cognition, our ability of imagine, resolve problems, make decisions, and put in order is lacking. If so, that would make the hypothesis as a theory of human cognition miserably incomplete. In this part, we explain how these fundamental human cognitive processes will account for the values. Awareness related to the use of common problem solving techniques is key biological awareness. There are no common trouble-solving techniques on hand to us that are teachable and learnable because we have developed to enhance such approaches as the most important knowledge. They can't educate people on how to use a change in the fact-end strategy, because all common people use the strategy immediately without guidance. Although primary biological knowledge is theoretically stored in an information stored, we do not actively obtain key knowledge using the usual concepts of information mentioned above. For instance, the idea of a functioning memory which is reduced in ability while dealing with new information, but which has no boundaries while working with long-term memory structured information do not have a place while dealing with predominant evolutionary awareness. When the thought processes and trouble-solving comprise key knowledge, the concepts of usual information processes become important when these procedures are extended to secondary knowledge. This might not be teachable to use a mean-end technique because we might be taught to use this naturally as part of our main biological knowledge. For example, using this tactic to a new scientific problem brings into play the concepts of general information processing structures, since mainly scientific knowledge is physically secondary. We may perhaps not require to actively process the methods of make sense to say-end analysis in operational memory due to the mechanisms of mean-end analysis were acquired as primary knowledge, but the less important information of new calculations in operational memory needs to be processed. While dealing with physically minor knowledge, the interrelationship between borrowing and restructuring, arbitrariness as genesis, and values of the ecological association and linking that to classify how cognitive method at a higher level function. When we retain information, long-term memory information is caught up in working memory in the first instance, using the concept of environmental organization and linking. Something new has been developed at this stage because we are simply placing the facts earlier learned into operational memory. Thinking needs one or two additional procedures, or together. Using the borrowing and reorganizing theory, we can reorganize the information just as information is restructured during sexual reproduction, substitute merging, or transposition. If we have not received supplementary information signifying how to reorganize the preceding information, then just as in the case of sexual reproduction, we need to restructure the information randomly before checking for usefulness. Does the latest, restructured information reach our environmental targets? If it performs, we may use the new restructured knowledge by storing it for later use in its current form in long-term memory or for further reorganization during thinking. Most learning can be imagined to happen through this restructuring cycle just like most variability in evolution. Fully new information is generated, on less frequent occasions. If information is moved from long-term to functioning memory using the classifying and connecting theory of the system rather than restructuring it, the information may be altered arbitrarily just as the metamorphosis changes the information in DNA. This spontaneous modification of long-term store knowledge is also a part of thought process. As is the case with restructuring, the effects of modifying the information arbitrarily may not be understood before the modification happens. The thinking procedure allows us to change the information spontaneously and then, as in the case of alteration, check the modification for usefulness with successful modifications in long-term memory being available for storage. We consider that these exchanges between borrowing and restructuring, arbitrariness as origin, and concepts of ecological organization and linkage form the base for elevated processes. When solving problems, either we restructure preceding information and check it for efficacy as it happens during sexual reproduction or arbitrarily create new information as it occurs during transformation, which also needs to be tested for validity. Failure to learn long-term memory indicates that randomization seems inevitable. Consider a trouble resolver trying to address a problem by solving analog problems. If we know how to solve the targeted problem via the source equivalent, the random generation and test component of the successful and the new information can be processed. Sometimes attempts to solve analog problems do not succeed because any of the correct analog is not used or is misused—if we do not have information about how to employ an analog, the process needs arbitrary creation and testing. We propose that these procedures might indicate why it can be so hard to reason by analogy. It must be noted that analysis by similarity involves concurrent planning of information on two or more problems which are likely to force a heavy work memory load. The equivalent mechanism allows for a viable decision-making sub-stratum. It can be concluded that all decisions rely on a mixture of preceding knowledge where that information is existing and arbitrary production and testing to the degree it is not obtainable. The way the intelligence used includes inaccurate information based on feelings or other variables and contributes to renowned, decision-based cognitive illusions. Planning is an activity characteristic of human action. This can be viewed as a specific example of consideration and can thus be studied using relations between borrowing and restructuring, arbitrariness as genesis, and concept of ecological organization and linking. We learn to make use of specific plans and specific preparation procedures. As with other cognitive tasks, most of our strategies are derived from the long-term memory of our own or of someone else. They can be restructured to suit a particular situation, but if so, to decide whether the reorganized plan is successful, they need an arbitrary creation and testing process. The random generation and testing of novel procedures will generate entirely new plans. The theory outlined here does not eradicate thought, but efforts to explain it in terms of borrowing and reorganization interactions, randomness as genesis and concepts of environmental organization and linking. Whether the theory is true may rely on moral rather than empirical issues, at least in some respects. For example, while the concept of randomness as genesis may be controversial in human cognition, we cannot explicitly check whether a person must invoke that principle rather than a choice in the lack of appropriate information because we are incapable to produce any functional substitute. If applicable information, including acquaintance that could allow us to simplify the similar circumstances, is inaccessible, there seems to be no substitute for random generation and testing to which we could develop an empiric test. No such solution appears to exist either in natural or artificial information processing systems. In comparison, we do learn that natural selection utilizes random generation and checking during evolution. As noted above, with regard to empirical concerns, as problem solvers approach complex, multi-move problems, they will often arrive at more dead ends than proper moves. The random generation and testing will explain several dead ends that an alternative would also need to explain.

4.9 Summary

The interpretation and reasoning approach to information-processing (IP) emerged as a response to behaviorism. This response concerned primarily the nature of scientific psychological explanation. The behavioral "standard" account, phrased in purely external terms, was substituted with a "realistic" account, articulated in words of internal institutions and processes. The study of the abstract vocabulary used in IP psychology indicates that the case is undisciplined. An analysis of the abstract vocabulary used in IP psychology reveals that a large number of terms are used simultaneously, there is no unambiguous level of analysis, and basic principles such as knowledge and retrieval remain largely unclear. Nevertheless, the IP method has evolved over the past 25 years into a complex and systematic experimental science. A glance at real action indicates the fundamental cause of his success. The approach is not so much concerned with the human information processor's absolute or intrinsic properties, but with what its relative or differential properties can be termed. In terms of the systematic vocabulary of a conceptual system, further study of this function of the IP methodology renders clear the reason for its success. The IP method can be seen as constructing an analytical discrepancy calculus on an undefined class of objects, phrased in terms of theoretical "theory-neutral" vocabulary of interpretation, and with operators structurally similar to logical operators. The reinterpretation of what the path to IP is all about offers a range of benefits. It reinforces its role as an autonomous discipline, clarifies its connection within the cognitive science community to other methods to psychology and other sciences, and renders it independent of methodological subtleties.

References

Abbot, B. (2002). *Human memory*. Fort Wayne, IN: Indiana University-Purdue University at Fort Wayne, Psychology Department. Retrieved June 22, 2002.

Anderson, M. (1998). *Jerome Bruner. "Educational psychology"*. Portland, OR: Cortland College. Retrieved June 12, 2002.

Atkinson, R., & Shiffrin, R. (1968). Human memory. A proposed system and its control processes. In K. W. Spence & J. T. Spence (Eds.), *The psychology of learning and motivation* (Vol. 2, pp. 89–195). New York, NY: Academic Press.

Clark, J. M., & Paivio, A. (1991). Dual coding theory and education. *Educational Psychology Review, 3*(3), 149–170.

Craik, F. I. M., & Lockhart, R. S. (1972). Levels of processing: A framework for memory research 1. *Journal of Verbal Learning and Verbal Behavior, 11*, 671–684.

Driscoll, M. (2001). *Psychology of learning for assessment* (2nd ed.). Boston, MA: Allyn & Bacon.

Eliasmith, C. (Ed.). (2001). *Memory. Dictionary of philosophy of mind*. Pullman, WA: Washington State University. Retrieved June 19, 2002.

Huitt, W. (2000). *The information processing approach. Educational psychology interactive*. Valdosta, GA: Valdosta State University. Retrieved June 12, 2002.

Kearsley, G. (2001a). Dual coding theory. In A. Paivio (Ed.), *Theory into practice*. Jacksonville, FL: Jacksonville State University. Retrieved June 14, 2002.

Kearsley, G. (2001b). *Levels of processing. Theory into practice*. Jacksonville, FL: Jacksonville State University. Retrieved June 14, 2002.

McClelland, J., & Rumelhart, D. (1981). An interactive activation model of context effects in letter perception. Part I: An account of basic findings. *Psychological Review, 88*, 375–407.

Mohanty, S. N., & Suar, D. (2013a). Decision-making in positive and negative prospects: Influence of certainty and affectivity. *International Journal of Advances in Psychology, 2*(1),19–28.

Mohanty, S. N. & Suar, D. (2013b). Influence of mood states, group discussion, and interpersonal comparison on change in decision-making and information processing. *International Journal of Decision Science, Risk and Management, 5(2)*,101–123.

O'reilly, T. (2005). Web 2.0: compact definition, 137–158.

Porter, A. L., & Read W. H., eds. (1998). The information revolution: Current and future consequences. Greenwood Publishing Group.

Rumelhart, D. (1980). *Schemata: The building blocks of cognition*. Hillsdale, NJ: Erlbaum.

Rumelhart, D. E., & McClelland J. L. (1986). On learning the past tenses of English verbs, 216–271.

Suthers, D. (1996). *Attention and automaticity*. Pittsburgh, PA: University of Pittsburg, Learning Research and Development Center. Retrieved June 21, 2002.

Wiener, N. (1948). Cybernetics or control and communication in the animal and the machine. Technology Press.

Winn, W., & Snyder, D. (2001). *Mental representation. The handbook of research for educational communications and technology (Chap. 5)*. Bloomington, IN: The Association of Educational Communications and Technology. Retrieved June 12, 2002.

Chapter 5
Effect on the Emotional Self-Esteem of Women with Reference to Make-Up

Richa Goel, Seema Sahai, and Gurinder Singh

Contents

5.1 Introduction

Makeup is considered efficient at rising judgments that the user might need to modulate, with people showing to be healthier and earning additional. Cosmetics additionally modify the behavior, particularly men, who are more influenced with external appearances. Men tip better, higher, and in more frequency to waitresses who wear makeup. Makeup is known to increase attractiveness, especially such that contributes very less to alter an individual's attractiveness within the population. We tend to look to how necessary it is when cosmetics influence attractiveness. Vanity, as a whole, created an impression on attire consumption, vesture perspective, and

R. Goel (✉)
Amity International Business School, Amity University, Noida, Uttar Pradesh, India
e-mail: rgoel@amity.edu

S. Sahai · G. Singh
Amity International Business School, Amity University Uttar Pradesh, Noida, India
e-mail: ssahai@amity.edu; gsingh@amity.edu

S. N. Mohanty (ed.), *Emotion and Information Processing*,
https://doi.org/10.1007/978-3-030-48849-9_5

look improvement. For this reason, vanity is additionally expected to have an effect on makeup consumption.

Makeup satisfaction could be a degree of satisfaction of makeup progress or result. The psychological effects of makeup could embrace caress of skin upon contact, completion of hope of transformation, and stress relief or pleasant tension throughout makeups. Moods are also elevated when make up is applied (Mohanty S N (2014))

Negative vanity, an individual who values his own worth and talent negatively, finds superficial satisfaction and fulfillment from makeups and so will increase makeup satisfaction as critical positive vanity, i.e., an individual who would value his own worth and talent absolutely. In different words, vanity seems to play a large role in makeup satisfaction. Cosmetics imparts the esteem that our general public places on women's physical appearance and furthermore serves to start young women into womanhood, acquainting, and furnishing methods with training congruity and inauthenticity, which for some women fill in as basic ingrained instincts all through their lifetime. Allowing what is encircled as an essential change of women, cosmetics can convey profoundly disturbing messages that mirror society's mentalities toward women. By using cosmetics and through other social practices, these states of mind are in the long run disguised, enabling the cosmetics business to gain by the specific uneasiness it made. No place is the cosmetics business' control of women as significant similarly as with the medicalization of the maturing procedure.

Usually, goods like compact, blush, mascaras, and lip products, to solely name many, are adopted by women to reinforce their look. Drawing around the eyes with a variety of colors and makeup to dark eyebrows and eyelashes was popular in ancient Egypt and Arabia. The residents of Egypt held thoughts powerfully within the use of cosmetics for mystical purposes. Their belief was that their look had an on-the-spot correlation to their spirituality. They used cosmetics not only to look exceptionally stunning but because they also believed that by painting around their eyes, they may get rid of evil spirits. As soon as the Greeks adopted the application of makeup, the medians of wearing the product modified majorly. Egyptians enhanced these products for the aim of spirituality, except for the Greeks wearing makeup was strictly for looks' sake. Centuries later, once the Roman Empire supported the employment of cosmetics, it created a high demand and utilization of the product all over Europe. Pumice stones were used to whiten teeth, remove callus, and rouge to stain cheeks, and a color was used to enhance lips. Back then, makeup was used not solely to boost looks however as an indication of rank. People who were labors and discovered within the fields forever had tanned skin; thus, it had been thought additional stunning to possess pale skin, an indication that you simply had cash and failed to go to work. Socially one would not be deemed to be a "nice"-looking woman if she wore several cosmetic products. The contemporary society teaches "Outward appearances are the foremost necessary assets for girls." In an exceeding society wherever bound facial appearances and figures are attributable as beauty, girls these days are investing heavily in their outward look. Additionally, young and old women seek more attention and approval from the environment by doing so. Makeups cowl facial flaws and enhance one's self-assurance and social relations. Individuals' vanity plays a major role in actions, motivations, achievements, and social relationships. The continuation of low vanity in look throughout this era could cause negative effects till adolescence and adulthood. Since the average age of makeup people has gone down, the scale of the marketplace for teenaged makeup product has matured. In the Far-Western society, girls deliver the bulk of self-acceptance and maybe the foremost current behavior of this sort is that the application of makeup. This behavior serves in favor of the cosmetic business which is now worth billions. Girls report a violation of application of cosmetics for a number of reasons, ranging from anxiety concerning external looks, conforming to social norms, and self-consciousness through to showing additional sociable and assertive to others. Within time makeup has had many alternative significances, shaping spirituality and even shaping people. One issue has forever stayed unchanged which is that the concept makeup causes one to look additional stunning.

While examining essential parts of life, finding an associate has reliably been seen as being of inconceivable essentialness. From the time adolescents are little, they are given that when they grow up they will one day meet someone, encounter enthusiastic affections for, and get hitched. It is sensible to express that when you meet that individual, you will find them appealing or alluring.

Research demonstrates that facially charming women are more capable to get hitched and will most likely marry of higher societal position. It has furthermore been found that facial magnificence is of more critical essentialness to men when picking an associate, than is for women while seeing men as their mate. It is imperative to look at the association between the purchasing and use of therapeutic things, in association with target acknowledgment and self-perspective of heavenliness. With the officially made reference to factors, it gives off an impression of being nothing sudden that billions of dollars a year are being spent by women on things to enhance themselves. Understanding greatness' effect on certainty is a troublesome topic (Mohanty & Suar, 2013a, b). There are a couple of examinations that exhibit a positive association among charm and certainty, more so for women than men, in any case a substantial number of the examinations seem to delineate weak associations between the two. Anyway, there are moreover analyses that decidedly propose an obvious association among brilliance and certainty using extents of mental thriving. In light of social effects, it is not surprising that frequently time's women need to look and show sureness, and furthermore, alluring quality. While these segments are so far dynamic, there is no denying that getting and applying enhancing things expect some activity in the attempted achievement of perfection. It is respected that individuals that are seen as superb are acknowledged to have higher certainty and assurance. It was moreover found that being seen as charming was related to drawing in quality which created more excitement from the opposite sex.

Social perception is a situation women have to accustom to and deal with worldwide. In America, for instance, the craving to be tan leads to numerous women to go to salons where they have to stay in heat that expands their risk of skin disease. The craving of individuals to comply with an admired, even inaccessible, standard of excellence can negatively affect both their physical and psychological well-being. If it is doable for particular brands of headway to incite an extension in therapeutic issues or a reducing in the reinforcing of a particular social event of people, this is a factor that must be considered in picking whether to grasp that particular enhancement framework. Right when what makes headway extremely supportable. Regardless, one need for prudent progression that can be settled upon is that plausible enhancement must benefit the people who do not generally have control in the general population eye, the people who are disparaged. In India, those social affairs reach out from lower standing people and Dalits to inborn get-togethers to women. The issue of eminence is particularly important while contemplating this last assembling, since women, when all is said in done, are the all-inclusive community most affected by brilliance models. India is an overall population encountering a passionate addition in its Westernization, not entirely caused by the opening up of its economy to all-inclusive associations, a phase taken by the organization with the true objective to propel the advancement and enhancement of the Indian economy. This Westernization has affected what associations work in India and also various parts of urban Indian culture. One needs to simply look at the streets of tremendous urban zones like Delhi or Jaipur and note how many individuals from the more energetic age are clad in jeans instead of salwar to perceive how on a very basic level this change has quite recently influenced Indian culture. Furthermore, with the modification in these parts of culture, it is nothing startling that rules of magnificence have every one of the reserves of being evolving as well. In any case, are these movements positive or negative? From different perspectives, women in Western culture appear to have more alternatives and openings: when in doubt, they can marry whom they pick, live self-rulingly, take part in sexual relations outside of marriage without being avoided, and wear revealing clothing. It is not sure whether these, and other, open entryways are all around empowering, anyway for the people who trust they are, it might make the feeling that Western culture is more captivating for women than Indian culture. For this, circumstance, it might be normal that grasping Western checks, for gloriousness, sexuality, or whatever else, would empower Indian women. This is a hazardous unsafe supposition, paying little mind to whether we completely trust the case that women are in general more liberated in Western culture than in Indian culture. For a specific something, paying little heed to whether an overall population when all is said and done treats women better than another, this does not suggest that all aspects of that society's culture is more helpful to women. As a matter of fact, the issue of brilliance and self-recognition is one a player in Western culture which fights with its treatment of women. Dietary issues are just a single of different locales in which this fight is appeared, as Western social requests generally have a basically higher rate of the diffuses than non-Western social requests. So there exists a strong likelihood that paying little heed to whether Westernization all around were to decidedly influence Indian women, Westernization of greatness benchmarks may at present have a negative one. So, what impacts has the Westernization of greatness standards truly had on Indian women? Since India is a culture with undeniably by and large extraordinary contemplations in regard to women's employments than those of most present day Western culture, it may be handy in India to perceive considerations of women that are intrinsic in Indian culture and musings that have been gotten from the Western world. Along these lines, it winds up possible to dissect definitely what influence globalization has had on Indian women and to set out upon the more troublesome task of examining whether this impact has been certain. To by and large understand this issue, there are four requests that must be answered.

The first is the thing that Indian checks of perfection were obviously, which must be offered a clarification to get a sensible picture of how they have changed. The second is the thing that Indian magnificence measures are right now. The third is the thing that effects these rules of radiance, and particularly the changes in these standards of brilliance, have on women in India. Additionally, the fourth request is whether these effects are sure ones.

Statement of the Problem:

- The purpose of the study was to determine whether makeup is directly effective on a woman's personal image of herself.
- To study if the effect of makeup satisfaction is direct on the self-esteem and confidence of women.

Questions of the Study

- Does not applying makeup make women self-conscious of their appearance?
- What is the effect on women who usually do not wear makeup, toward their own appearance when they apply it?
- Do women consider themselves less desirable if they do not apply makeup?
- Is makeup a necessity or a societal norm?

5.2 Objectives of the Study

- To determine how makeup affects the self-esteem of an individual.
- To understand if females feel more confident if they regularly use makeup.
- To determine if makeup augments one's identity and enhance their self-esteem.
- To understand the factors contributing to the consumer psychology of purchasing and applying makeup.

5.3 Literature Review

Davies (2016) aims to look at how women are presently pictured in restorative promoting, the impact these adverts wear feminine attitudes and thus the effect this has on shopper getting behavior. This analysis primarily seeks to grasp the mental feelings felt by shoppers concerning cosmetic ads and successively; however, this impacts the merchandize that they obtain. It will analyze how advertisements specialize in psychographic factors, examining how they fight and uphold feelings like confidence, social acknowledgment, and increasing vanity. It will analyze the assurances these cosmetic corporations build to their shoppers, concerning how the merchandize can "improve" their physical look in how this impacts shopper getting behavior therefore. Hemapatil and Bakkappa (2012) describe culture and utilization have a new relationship within the modern times. The speedily increasing significance of international selling has diode marketers, specialists, and academicians alike, to hunt a more profound understanding of how shoppers and markets dissent round the globe. The planet economy is turning into a progressive society. Consumption choices created within the market cannot be viewed as a freelance event—these are closely connected with values and social relationship and social dependability. Cosmetics are an omnipresent part of women's client culture one among the imperatives of up to date life, and represent one among the foremost vital ways that girls gift and remodel their open persona. In today's society, excellence and physical attractiveness are perpetually stressed as fascinating and splendid attributes. This analysis focuses on consumers' cosmetics shopping for deciding process because of culture. Isabel and de Nadai (2016) outlined women's defrayal comparatively additional on beauty product throughout economic recessions. The aim of this analysis is to know the underlying motives leading women to shift their shopping priorities throughout economic downturns. The paper's findings counsel that the two main factors driving the lipstick impact are: (1) women looking out to feel happier with their overall personal life (i.e., overall level of happiness, sociableness, and self-esteem) and (2) women looking out to feel additional assured in their skilled lives.

Gelles's (2011) investigation seeks to look at specifically what impact this globalized media has on its watchers and, especially, on their views on beauty. This study can scrutinize specifically what impacts these models of beauty wear the women being estimated against them. Here they fight to outline an objective customary for excellence, to classify specifically what makes somebody lovely in the eyes of others. These investigations have examined problems as variation as midriff-to-hip or abdomen-to-chest quantitative relation and facial symmetry. Several of those studies offer compelling proof for his or her claims. However, though these factors add to beauty, it might be troublesome to say that they are the main real determiners of what constitutes a gorgeous look. This leaves several areas within which determination of magnificence is abstract, and shifts cross-culturally. For instance, simply in civilization, think about the development of how society sees on the best skin tone. Tagai, Ohtaka, and Nittono (2016) say numerous women wear facial makeup to intensify their charm and engaging quality. Makeup could vary from common (light) to exciting (heavy), relying of the setting of social things, a stress on current social makeup patterns. This investigation examined how light-weight makeup and significant makeup impacted engaging quality appraisals and identity verification. The current study recommends that light-weight makeup is desirable to significant makeup in this light-weight makeup does not interfere with individual recognition and gives holders positive impressions. Ohno (2005a) describes cosmetics usage, as associate degree alterable side of physical attractiveness and individual look, has the potential to own goodish impact on associate degree of individual's body image, self-perceived attractiveness, and vanity. The aim of this study was to spot the relationships between the variables of cosmetics usage, self-perceived attractiveness (facial satisfaction), body image, and vanity in college-age females. In this paper, a significant relation between body image and perception of external appearance with respect to cosmetic usage was found. Certain relationships between facial appearance and satisfaction along with self-esteem and body image was also present in the paper.

Nishihara et al. (2013) shows that the motivation behind this investigation is to make it clear what measure the consequences of makeup, fragrance, and skin care product on the grouping, and psychological state of females. They studied Japanese women of the ages 18–30. Makeup, fragrance, and skin care products can be said to be associated with higher health for school kids. In addition, hair care could facilitate and maintain psychological state in female students. Regular usage of makeup, fragrance, and skin care product can be implicit to become a precise reason why students perform higher and even have a precise routine to stay them union. Jones and Kramer (2016) say that society operates as if it has the potential to understand a woman from her physical look, notably from her face. It is assumed that one will understand a woman's health, competence, sexuality, and mental state just by staring at her face. Ohno (2005b) states that using the objectification theory, i.e., the female body is an object to be identified, appreciated, and evaluated upon. The usage of cosmetics is considered to be a method to enhance one's appearance, but temporarily. Due to particular standards, women tie their body image and certain external appearances to their self-esteem and confidence in order to meet them. This creates a situation where women who feel positive about their appearance portray positive aspects when it comes to body image, self-confidence, and self-esteem. Whereas, when certain women are negatively associated with the same, they feel dejected and depressed.

Britton (2012) mentions there has been plenty of analysis done on the impact that promoting within the kind and excellence business has on women. By creating commercials with unlikely photos of magnificence, it has caused tension, low confidence, and low fearlessness in various women. The larger a part of these negative feelings originates from misery among body and look. Less analysis has been performed distinguishing with trending merchandize and the way this will have a sway on women, and the way women will utilize beautifiers to regulate their look. This paper talks about this analysis that spotlights on the corrective business' impact on women.

Silverio (2010) shows that in each general public there is a standard of appearance that the populace is relied upon to pursue. There are those that endeavor outside of the crate and push the envelope as it might; however, there are general rules that most appear to concur are sensible, inside the setting of their general public. In many societies, these limits of conduct are viewed as social standards. It has frequently been a suspicion in the present culture that on the off chance that one does not set aside the opportunity to prepare themselves appropriately, there is some kind of problem with them. Commonly the individuals who do not stay aware of these prepping propensities are accepted to have a psychological illness or imperfection, be inadequately thought about, or have a low assessment of themselves. It has likewise turned into a social standard in the United States for women to wear cosmetics. It is an acknowledged procedure that women ought to wake up early in the day and, alongside conventional preparing propensities, apply items to their face to mask flawed territories and upgrade delightful ones. The purpose of this paper was to decide if wearing cosmetics specifically influences a woman's perception of beauty as seen by her. Certain questions come up in the paper which comply with the paper's objective. They focus on if makeup makes a woman feel confident, more attractive, special, more viewed, etc. Komulainen and Hjort (2017) state that the point of this examination was to reveal how sexual orientation transgressing hones are performed in male excellence vloggers' cosmetics utilization and to contribute into the comprehension of how men are developing their sex personality when entering female utilization spaces. Situated inside the purchaser culture hypothesis field, the examination means to grow current viewpoints on gendered utilization on a hypothetical level. Jeffreestar, Manny Mua, PatrickStarrr, James Charles, and Gabriel Zamora were incorporated into the paper for research. This explore sought after to scrutinize this end inside the setting of male magnificence vloggers' cosmetics utilization. The specialists watched, deciphered, and broke down 100 YouTube recordings from five distinctive male-distinguished excellence vloggers. The discoveries proposed that the cosmetics utilization in these recordings deduce to a liquid hybridization of sex develops. The vloggers depicted both extremely manly and feminine attributes in the recordings and blended them in a free, liquid way. The cosmetics utilization conveys implications past the quest for magnificence for the vloggers, it goes about as a method for prominent self-introduction, experimentation, and imagination. The vloggers depict a message of decent variety, inclusivity and equity inside the setting of the customarily exceptionally gendered utilization circle of cosmetics. Klein (2013) describes that the inspiration driving this paper is to grasp and blame the activity of online life in the enhancement and additionally support of dietary issues, scattered eating, and body disillusionment in school developed women. School women are exceptionally exposed against the impact that electronic life can have on their self-observation as they develop a point of view toward their bodies and recognize the developmental changes that occurred in the midst of immaturity. This paper gives evidence that there is an association between the continuous surge in befuddled eating and high use of web-based life. This examination overviews the conduct by which the thin flawless as portrayed in publicizing urges women to look a particular, improbable way.

Verbickaite investigates the effect that beautifying agents have on female prosperity with respect to mental, physical, social, vocation, and training perspectives. In the western world, corrective items are acknowledged, as well as regularly urged to be utilized for Caucasian females to look most alluring, subsequently be seen all the more decidedly by society. Through different media channels, the impression of excellence are formed and stereotyped. Through such channels, females find out about appearance standards. As they gain from them, females feel compelled to utilize beautifiers with the end goal to control and improve their face highlights to coordinate the desires. Through this investigation, it was discovered that females are vigorously subject to the use of makeup with the end goal to feel sure and agreeable in social circumstances. Female members of the survey were less satisfied about their regular appearance contrasted with when utilizing beauty care products, dominant part of them utilize beautifying agents in every case then they are in broad daylight. With the utilization of beauty care products, females can seem more social, secure, sure, or prevalent. Females wearing beautifiers are seen as more feminine, in this way in applying for occupations viewed as manly cosmetics can be disadvantageous. Beautifying agents can likewise influence impression of female's social class and well-being. Narang (2013) describes that the inspiration driving this examination is to consider how beautifying agents impacts the way in which outcasts see particular character and physical characteristics. Past research has found that low certainty, low conviction, and anomalous measures of pressure provoke extended beauty care products utilize; and extended beautifiers utilize firmly partners to drawing in quality, womanliness, and provocativeness and antagonistically relates to pleasantness, moral quality, and capacity. Untouchable observers' impression of beautifying agents is affected by the going with factors: situation, race, and time. The present research may prompt women on the most capable strategy to tailor their beauty care products choices to affect particular perceptions. Also, specialists and other social protection specialists may use this information when diagnosing or treating a patient. Associations and endeavors may use this figuring out how to diminish tendency.

Tagai, Ohtaka, and Nittono (2016) state that numerous women wear cosmetics to emphasize their allure and engaging quality. Makeup may change from normal (light) to charming (generous), depending of the setting of social conditions, a complement on propriety, and current societal beautifiers designs. This investigation inspected how light cosmetics and overwhelming cosmetics impacted engaging quality appraisals and facial acknowledgment. In a rating undertaking, 38 Japanese women assigned drawing in quality examinations to 36 Japanese female appearances without any beauty care products, light makeup, and overpowering beautifiers (12 each). In an ensuing affirmation errand, the individuals were given 36 old and 36 new faces. Results demonstrated that appeal was assessed most hoisted for the light beauty care products countenances and minimum for the no beautifying agent faces. Strikingly, affirmation execution was higher for the no makeup and light show up than for the mind-boggling beautifying agents faces. Faces with overpowering beautifying agents conveyed a higher rate of false affirmation than did diverse appearances, maybe in light of the fact that generous beauty care products establish a connection of the style of beautifiers itself, instead of the individual wearing the makeup. The present examination recommends that light beauty care products are attractive over considerable beautifying agents in that light beautifiers do not intrude with individual affirmation and give onlookers positive impressions.

Lee and Oh (2018) describe cosmetics covers facial defects and improves one's fearlessness and relational relations. In addition to the fact that it serves as dress for the face, cosmetics has turned into a medium that associates human feeling. This investigation directed an examination on fifth and sixth grade young women to break down cosmetics with the goal of appearance upgrade. At the end of the day, the reason for the examination is to clarify how mental attributes may influence cosmetics contribution and fulfilment.

Davis (2013) mentions that the inspiration driving this examination is to research African American women's demeanors toward excellence care items, how women use cosmetics, and how usage of the things is related to African American women's self-character and self-impression of appearance. The precedent was 18 African American women who were inhabitants of Pine Bluff, Arkansas, or Ames, Iowa. The women began wearing beautifiers at different periods in their lives in the midst of youth, or energy. Searching for excellence care items could be an issue for a couple of individuals. Some declared inconveniences in finding the right foundation to organize their skin creation. Various individuals assumed that finding magnificence care items was an all-in or full-scale encounter when it came to searching for more affordable cosmetics in neighborhood stores. Working with a specialist at a retail foundation counter improved the experience. Stores in the commonly White Iowa region passed on couple of cosmetics appropriate for African American women's skin tones, making much disillusionment with deference the respondents around there. Gigantic quantities of the women spent more money on quality beautifiers that they felt enhanced their look and caused less skin issues. An impressive part of the individuals supported a comparable brand of higher-end excellence care items. The women adequately chase down YouTube instructional activities to learn new devices and techniques on the most capable strategy to apply their magnificence care items. For an impressive part of the women, wearing beautifiers is a sort of "charm" that enhanced their gloriousness and valor. Countless women illuminated that applying embellishing operators was relied upon to enhance their features; for a few, beauty care products were seen as futile, yet a better than average extension. Each and every women used restorative to demonstrate a romanticized appearance and had a great time wearing them.

Dickman (2010) demonstrates that the examination led to explore if there could be a connection between self-perception and the utilization of beauty care products. The exploration looked for first to recognize the self-perception and beauty care product utilization among the objective populace, and after that to look for conceivable connections between self-perception and utilization factors. An establishment for the examination was spread out by presenting the Finnish beautifiers showcase and an assortment of self-perception angles. The hypothetical system comprised of the most part of past research discoveries inside the territory of self-perception and beautifying agents. The experimental research was led by utilizing a quantitative strategy. The example outlined comprised of mid-range group of years old women understudies in the more prominent Helsinki district. Information was gathered for a reason for getting numerical information for factual and connection investigation. Factual measures were utilized to recognize the self-perception and beautifier utilization propensities among the respondents. Connection measures were utilized to investigate a conceivable connection between self-perception and makeup utilization. The examination results showed that a feeble connection between a few parts of self-perception and makeup utilization was found. Utilization was commanded by adapting conduct, which was communicated through makeup use sums and interest in appearance administration. Likewise, a buyer profile for both self-perception and beautifiers utilization was produced dependent on factual discoveries. Brinegar and Weddle (2014) plan to inspect the connection between cosmetic usage and consumption and confidence. Old research exhibits that appearance impacts all things considered success and that beautifiers impact appearance; thus, beauty care products may affect as a rule thriving. Researchers foreseen that subjects who had cut down certainty would use more beautifying agents simply more a significant part of the time. To examine this hypothesis subjects completed a certainty survey sought after by a beautifying agents review. While subjects were completing the surveys, experts requested the level of beautifiers the subjects were wearing. The data assembled exhibited a positive association between the subjects' report of beautifiers that they were at present wearing and the proportion of beautifying agents the examiners evaluated them as wearing. The results furthermore showed a positive connection between the beauty care products that the subjects were at present wearing and the level of physical certainty. This indicated subjects with lower certainty were most likely going to wear more beautifiers. There was no association between social certainty and beauty care products utilize, nor was there a connection between various arrangements of beautifying agents utilize and certainty. This suggests the reason that affects association among beauty care products and certainty cannot be settled, and future research should be controlled to choose this relationship.

Sankaranarayanan (2017) states that school going young girls understudies are one of the principle focuses of corrective organizations. Contrasts happen among these understudies as far as utilization, the sum spent, benefits looked for, brands favored and dispositions held. It will be useful and enthusiasm to the two scholastics and advertisers to investigate whether diverse sections exist in this market and the qualities of these fragments. An endeavor is made in this examination to portion this market utilizing use, state of mind and advantage recognition. An example of 120 school going girls understudies have been studied utilizing a poll, and the information have been examined utilizing bunch investigation. This brought about three groups with varying use, state of mind, and advantage discernments (free non-adherents, heavy advantage searchers, and ward periodically). Cross organization of profiling factors with bunch participation and chi-square examination uncovered no noteworthy contrasts in statistic profile, conduct attributes, and brand inclinations crosswise over groups. The sections were found to contrast as far as their normal going through on beauty care products with intermittent being the most astounding spenders pursued by substantial advantage searchers and free non-adherents being the least spenders.

Baxter, Hastings, Law, and Glass (2008) describe that various American women encounter the evil impacts of poor self-discernment and low dauntlessness. Wearing makeup is something they can do to quickly and unexpectedly change their appearance, in this way growing assurance. The present examination is an examination of the association between magnificence care items, their match to particular conditions, and the consequent anxiety levels. Apprehension was used as an extent of assurance. Individuals changed their beautifying agents in three exceptional styles in two certifiable conditions. After each situation, the individuals adjusted a short report about their feelings about the beautifying agents and the condition they were in.

5.4 Conceptual Framework

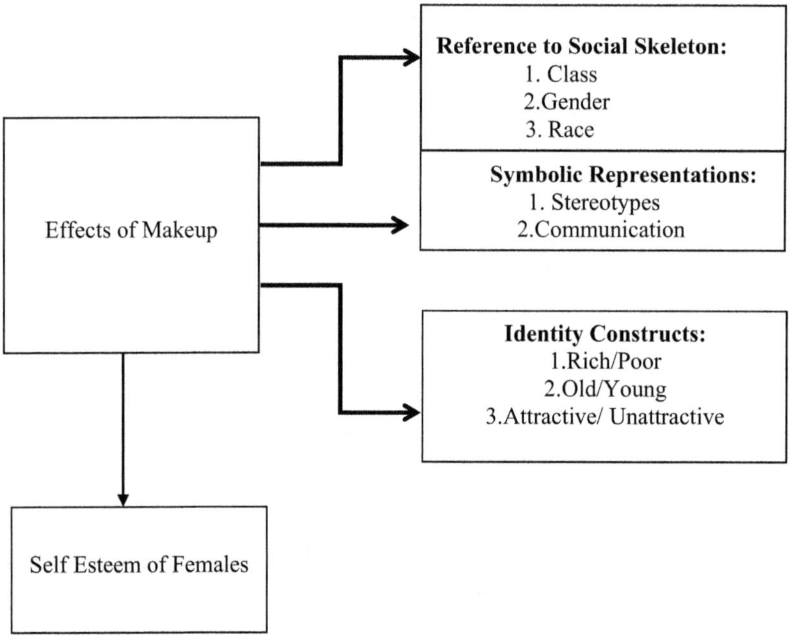

5.5 Hypothesis

- There is a significant relationship between one's income and choice of makeup brand.
- There is a significant relation between increase in self-esteem while wearing makeup.
- There is a significant relationship between societal pressure and application of makeup.
- There is a significant relation between age affecting choice in makeup.

5.6 Methodology

The research conducted has used primary and secondary data. The type of research conducted was exploratory in nature. A sample size of 200 respondents was taken out of which 196 responded. The data was collected and analyzed using SPSS software. Secondary sources were also used with respect to articles, research papers, etc. Descriptive statistics was done using mean, standard deviation, and frequency. Inferential statistics was obtained using correlation, reliability, and Cronbach's alpha.

5.7 Data Analysis

5.7.1 Finding out the Response Rate (Table 5.1)

Table 5.1 The number of questionnaires and their validity received for analysis

Responses	Frequency
Returned complete questionnaires	196
Returned incomplete	4
Total	200

5.7.2 Finding out Descriptive Statistics

5.7.2.1 Frequencies (Table 5.2)

Table 5.2 Socio-demographic information

Items		Frequency
Gender	Male	10
	Female	184
	Other	2
	Total	196
Age group	>18 years	3
	18–25 years	97
	25–40 years	49
	40–55 years	45
	Above 55 years	2
	Total	196
Education	10th	4
	12th	12
	Graduate	76
	Postgraduate	99
	Other	5
	Total	196
Occupation	Unemployed	28
	Employed	86
	Student	60
	Other	22
	Total	196
Income	Below 20k	94
	20–40k	40
	40–60k	16
	More than 60k	46
	Total	196
Marital status	Married	92
	Unmarried	104
	Total	196

5.7.3 Reliability

Table 5.3 shows the degree of the questionnaire's reliability. The value of alpha is 0.883 which is more than 0.6. This shows that the data is reliable for further analysis.

5.7.4 Using Mean and Standard Deviation

As we can see, the mean in Table 5.4 lies between 1 and 3, and it shows that the respondents have a positive approach, i.e., they opted to agree toward the parameters. Standard deviation shows the degree at which the mean is deviating from the actual mean.

5.7.5 Hypothesis 1

H0: There is no significant relationship between one's income and choice of makeup brand.
H1: There is a significant relationship between one's income and choice of makeup brand.

5.7.6 Using Correlation and Regression Analysis

Table 5.5 shows the relationship between self-esteem and income at 95% significance level.

5.7.7 Using Regression to Prove Hypothesis

Table 5.6 shows the significant relationship between self-esteem and income, where self-esteem is a dependent variable, and income is an independent variable. The equation is Self-esteem = 0.608 + Income * 0.355.

The test shows that there is a significant relationship between the level of self-esteem and the income of the individual.

So, we will accept H1 and reject H0.

Table 5.3 Reliability statistics

Cronbach's alpha	*N* of items
0.883	52

Table 5.4 Mean and standard deviation of the variables

Variables	Mean	Std. deviation	Variables	Mean	Std. deviation
Demo1	1.02	0.123	A14	3.49	1.237
Demo2	1.74	0.840	A15	3.50	1.238
Demo3	2.09	1.231	A16	3.06	1.342
Demo4	3.46	0.704	A17	3.08	1.213
Demo5	1.46	0.500	A18	2.85	1.270
Demo6	2.40	0.920	A19	3.31	1.355
S1	2.18	1.035	A20	1.37	0.485
S2	3.44	1.652	B1	1.38	0.486
S3	2.88	1.311	B2	1.61	0.488
S4	2.68	1.429	Age1	1.19	0.396
S5	4.97	1.370	Age2	1.19	0.396
S6	1.29	0.616	C1	1.70	0.458
S7	5.57	2.954	C2	1.71	0.455
A1	2.93	1.236	C3	1.33	0.470
A2	3.33	1.268	C4	2.97	1.362
A3	2.54	1.500	D1	1.39	0.488
A4	3.82	1.298	D2	1.39	0.488
A5	3.96	1.252	E1	3.07	1.224
A6	2.57	1.344	E2	1.26	0.440
A7	3.23	1.182	F1	1.28	0.450
A8	2.45	1.408	F2	2.06	1.346
A9	2.88	1.452	G1	1.17	0.380
A10	2.63	1.268	G2	1.22	0.415
A11	2.81	1.393	G3	1.17	0.380
A12	2.75	1.416	G4	3.15	1.765
A13	2.67	1.406	Self-esteem	1.14	0.346

Total number of respondents (N) = 196

Table 5.5 Correlation between self-esteem and income

Correlations			
		Self-esteem	*B*
Self-esteem	Pearson correlation	1	0.403**
	Sig. (two-tailed)		0.000
	N	196	196
B	Pearson correlation	0.403**	1
	Sig. (two-tailed)	0.000	
	N	196	196

Table 5.6 Regression between self-esteem and income

Coefficients[a]						
		Unstandardized coefficients		Standardized coefficients		
Model		*B*	Std. error	Beta	*t*	Sig.
1	(Constant)	0.608	0.089		6.791	0.000
	B	0.355	0.055	0.403	6.125	0.000

[a]Dependent variable: self-esteem

Table 5.7 Correlation between self-esteem and appearance

Correlations			
		Self-esteem	*A*
Self-esteem	Pearson correlation	1	0.178*
	Sig. (two-tailed)		0.013
	N	196	196
A	Pearson correlation	0.178*	1
	Sig. (two-tailed)	0.013	
	N	196	196

5.7.8 Hypothesis 2

H0: There is a no significant relationship between increase in self-esteem while wearing makeup.

H2: There is a significant relationship between increase in self-esteem while wearing makeup.

5.7.9 Using Correlation and Regression Analysis

Table 5.7 shows the relationship between self-esteem and appearance at 95% significance level.

5.7.10 Using Regression to Prove Hypothesis

The relationship between self-esteem and appearance after wearing makeup are shown in Tables 5.8 and 5.9, where self-esteem is a dependent variable, and appearance is an independent variable. It is represented by the equation: Self-esteem = 0.938 + appearance * 0.068.

The test shows that there is no significant relationship between the self-esteem after application of makeup.

So, we will accept H0 and reject H2.

5.7.11 Hypothesis 3

H0: There is no significant relationship between societal pressure and application of makeup.
H3: There is a significant relationship between societal pressure and application of makeup.

5.7.12 Using Correlation Analysis

Table 5.10 shows the relationship between self-esteem and the societal pressure for the application of makeup at 95% significance level. The relationship is positive in nature.

5.7.13 Using Regression to Prove Hypothesis

The test shows that there is a significant relationship between self-esteem and the societal pressure to apply makeup. It is represented by the equation: Self-esteem = 0.622 + appearance * 0.268 (Tables 5.11 and 5.12).

So, we will reject H0 and accept H3.

5.7.14 Hypothesis 4

H0: There is no significant relation between age affecting choice in makeup.
H4: There is a significant relation between age affecting choice in makeup.

5.7.15 Using Correlation and Regression Analysis

Table 5.13 shows the relationship between self-esteem and the influence of age in choice of makeup at 95% level of significance.

Table 5.8 Regression between self-esteem and appearance

ANOVA[a]						
Model		Sum of squares	Df	Mean square	*F*	Sig.
1	Regression	0.736	1	736	6.330	0.000[b]
	Residual	22.545	194	0.116		
	Total	23.281	195			

[a]Dependent variable: self-esteem
[b]Predictors: (Constant), *A*

Table 5.9 Regression equation depicted by the coefficients table

Coefficients[a]						
		Unstandardized coefficients		Standardized coefficients		
Model		*B*	Std. error	Beta	*t*	Sig.
1	(Constant)	0.938	0.083		11.269	0.000
	A	0.068	0.057	0.178	2.516	0.000

[a]Dependent variable: self-esteem

Table 5.10 Correlation between self-esteem and societal pressure

Correlations			
		Self-esteem	*C*
Self-esteem	Pearson correlation	1	0.322**
	Sig. (two-tailed)		0.000
	N	196	196
C	Pearson correlation	0.322**	1
	Sig. (two-tailed)	0.0000	
	N	196	196

Table 5.11 Regression of self-esteem and societal pressure

ANOVA[a]						
Model		Sum of squares	df	Mean square	*F*	Sig.
1	Regression	2.420	1	2.420	22.510	0.000[b]
	Residual	20.860	194	0.108		
	Total	23.281	195			

[a]Dependent variable: self-esteem
[b]Predictors: (Constant), *C*

Table 5.12 Regression equation depicted by the coefficients table

Coefficients[a]

Model		Unstandardized coefficients		Standardized coefficients		
		B	Std. error	Beta	*t*	Sig.
1	(Constant)	0.622	0.111		5.592	0.000
	C	0.268	0.056	0.322	4.744	0.000

[a]Dependent variable: self-esteem

Table 5.13 Correlation between self-esteem and influence of age

Correlations

		Self-esteem	Age
Self-esteem	Pearson correlation	1	0.237**
	Sig. (two-tailed)		0.001
	N	196	196
Age	Pearson correlation	0.237**	1
	Sig. (two-tailed)	0.001	
	N	196	196

Table 5.14 Regression of self-esteem and influence of age

ANOVA[a]

Model		Sum of squares	df	Mean square	*F*	Sig.
	Regression	1.312	1	1.312	11.585	0.001[b]
	Residual	21.969	194	0.113		
	Total	23.281	195			

[a]Dependent variable: self-esteem
[b]Predictors: (Constant), Age

5.7.16 *Using Regression to Prove Hypothesis*

Tables 5.14 and 5.15 show the relationship between self-esteem and influence of age.

The test shows that there is a significant relationship between the self-esteem and age of an individual during the application of makeup. It is represented by the equation: Self-esteem = 0.840 + age * 0.249.

So, we will reject H0 and accept H4.

Table 5.15 Regression equation depicted by the coefficients table

Coefficients[a]						
		Unstandardized coefficients		Standardized coefficients		
Model		*B*	Std. error	Beta	*t*	Sig.
	((Constant)	0.840	0.091		9.271	0.000
	Age	0.2490	0.073	0.237	3.401	0.001

[a]Dependent variable: Self-Esteem

5.8 Findings

This study was conducted to understand the effect of makeup on the self-esteem of women. A survey was conducted to understand the effect of makeup on self-esteem and various other factors which contribute to the same.

The findings of the study are as follows:

- Makeup does have an effect on the appearance and self-esteem of individuals who apply it.
- Makeup usage is strongly influenced and associated with the society and stereotypical factors.
- The usage of makeup is also a personal choice of many women.
- Makeup is a form of communication and a method to express one's personality in a more effective and artistic manner.

5.9 Conclusion

Makeup communicates the value that our society places on women's physical appearance and serves to initiate young girls into womanhood, introducing and providing a means to practice conformity and inauthenticity, which for many women serve as critical survival skills throughout their lifetime. As enabling what is framed as a necessary transformation of women, makeup stems from and reinforces deeply troubling messages that reflect society's attitudes toward women. Through makeup use and other cultural practices, these attitudes are eventually internalized, allowing the makeup industry to capitalize on the very anxiety it helped create. This study's main focus is on an issue which has been around for a very long period of time. The use of makeup is a personal choice or a societal norm for women to adopt. Thus, I would like to conclude, keeping my survey and its results in mind, that the usage of makeup and cosmetic procedures is known to give a significant boost to the self-esteem and confidence of a women.

5.10 Limitations

The study was carried out among females of various age groups. Certain limitations were observed during the survey. The survey was to be of 200 in total. One hundred ninety-six responses were received. Since the survey was an online-based survey, women who are active on social media were our respondents. This leaves a large subset of individuals who are less tech friendly. Thus, the study is mainly focused on people who are upper middle class in nature, use makeup majorly, and also consider appearance important in their opinion. Limited choice of answers were provided to the respondents for analyzing purposes. Inaccurate knowledge of concepts like self-esteem and confidence may have be considered similar. Thus, inaccurate answers may have been given.

5.11 Recommendations

Makeup has been used for many years to enhance one's appearance. The study focuses on how it is a significant aspect of a female's perception herself in the society and for herself. The need for makeup is very much reflecting on the society and the women. It has been a societal norm for a very long time to look beautiful to fit in. Women are asked to take care of their skin and their appearance for their marriages, and they start young to pluck their eyebrows and remove facial hair to look more attractive. The need to look prettier has always been an influence by the society. Though times have been changing majorly, in aspects of women wanting to look better for themselves. There have been significant studies where women are now embracing lighter and no-makeup looks to look natural and more appealing in the comfort of being themselves. Nowadays, women prefer thicker, bushier eyebrows for a more laid back look. The embracement of natural beauty is necessary for women.

The usage of makeup being optional and especially a personal choice is needed, since times and women are changing. Therefore, my final recommendations would be that women should not just rely on makeup entirely to feel good about themselves but also rely on their skills and be comfortable in their own skin. Along with comfort, women should not feel pressurized or bound to use makeup just because of the society being crucial on how they appear.

References

Baxter, R., Hastings, N., Law, A., & Glass, E. J. (2008). Influence of cosmetics on the confidence of college women: An exploratory study. *Animal Genetics, 39*(5), 561–563.

Brinegar, K., & Weddle, E. (2014). The correlation between makeup usage and self-esteem. *Full Thesis, 344*(1), 1–16. Retrieved from http://vault.hanover.edu/~altermattw/courses/344/papers/2014/brinegarweddle.pdf

Britton, A. M. (2012). University of new Hampshire scholars' repository the beauty industry's influence on women in society. *The beauty industry's influence on women in society*. Retrieved from http://scholars.unh.edu/honors%0a, http://scholars.unh.edu/honors/86

Davies, J. (2016, April). *Exploring how cosmetic advertising portrays women today and what impact it has on consumer buyer behaviour* (pp. 1–100).

Davis, L. (2013). *African American women's use of cosmetics products in relation to their attitudes and self-identity.* Graduate theses and dissertations, Iowa State University. Retrieved from http://lib.dr.iastate.edu/etd/13208

Dickman, C. (2010). Body image affects on cosmetics consumption among young females in the greater Helsinki region title: Body image affects on cosmetics consumption among young females in the greater Helsinki region supervisor (arcada): sveinn eldon. Retrieved. https://pdfs.semanticscholar.org/ca30/e92d166999c1f6d479a7a05cb67808a79c4d.pdf

Gelles, R. (2011). *Fair and lovely: Standards of beauty, globalization, and the modern Indian woman*. Independent Study Project (ISP).

Hemapatil, M., & Bakkappa, B. (2012). The influence of culture on cosmetics consumer behaviour. *IOSR Journal of Business and Management, 3*(4), 41–47. Retrieved from http://www.iosrjournals.org/iosr-jbm/papers/vol3-issue4/f0344147.pdf

Isabel, M., & De Nadai, M. (2016). *The lipstick effect and the self-enhancement motives.*

Jones, A. L., & Kramer, R. S. S. (2016). Facial cosmetics and attractiveness: Comparing the effect sizes of professionally-applied cosmetics and identity. *Plus one, 11*(10), 1–17. https://doi.org/10.1371/journal.pone.0164218

Klein, K. M. (2013). Why don't I look like her? *The impact of social media on female body image*. Claremont colleges. Retrieved from http://scholarship.claremont.edu/cgi/viewcontent.cgi?Article=1749&context=cmc_theses

Komulainen, V. L. U. & Hjort, M. L. U. (2017). BUSN39 2017, Department of Business Administration Men in Makeup: Performances of gender transgression in male beauty vloggers' makeup consumption, Student Paper

Lee, H., & Oh, H. (2018). The effects of self-esteem on makeup involvement and makeup satisfaction among elementary students. *Archives of Design Research, 31*(2), 87–95. https://doi.org/10.15187/adr.2018.05.31.2.87

Mohanty, S. N., Pratihar, D. K., & Suar, D. (2015). Study on influence of mood states on information processing during decision making using fuzzy reasoning tool and neuro-fuzzy system developed based on Mamdani approach. *International Journal of Fuzzy Computing Modeling, 1*(3), 252–268. https://doi.org/10.150.4/IJFCM.2015.069930

Mohanty, S. N., & Suar, D. (2013a). Decision-making in positive and negative prospects: Influence of certainty and affectivity. *International Journal of Advances in Psychology, 2*(1), 19–28.

Mohanty, S. N., & Suar, D. (2013b). Influence of mood states, group discussion, and interpersonal comparison on change in decision-making and information processing. *International Journal of Decision Science, Risk and Management, 5*(2), 101–123. https://doi.org/10.1504/IJDSRM.2013.058915

Mohanty, S. N., & Suar, D. (2014). Influence of Mood States on Decision Making under Uncertainty and Information processing. *Psychological Reports, 115*(1), 44–64.

Narang, D. (2013). The psychological factors that affect makeup usage and the perception of makeup in different situations. In: *Proceedings of the National Conference on Undergraduate Research* (pp. 915–918).

Nishihara, R., Wada, K., Akimitsu, O., Krejci, M., Noji, T., Nakade, M., … Harada, T. (2013). Effects of makeup, perfume and skincare product usage and hair care regimen on circadian typology, sleep habits and mental health in female Japanese students aged 18–30. *Psychology, 4*(3), 183–188. https://doi.org/10.4236/psych.2013.43028

Ohno, M. (2005a). Putting your best face forward: Relationships between cosmetics use, self-esteem, body image, and self-perceived attractiveness. *Physical Review B, 72*(10), 1–13.

Ohno, M. Z. L. (2005b). Making women: A look at the darker messages of cosmetics. *Physical Review B, 72*(10), 1–13.

Sankaranarayanan, K. G. (2017). Consumption pattern of cosmetics among female college students: A cluster analytic segmentation approach. *International Research Journal of Commerce, Arts, and Science, 8*(4), 119–133.

Silverio, L. (2010). Makeup's Effects on Self-Perception. OTS Master's Level Projects & Papers. 49. https://digitalcommons.odu.edu/ots_masters_projects/49

Tagai, K., Ohtaka, H., & Nittono, H. (2016). Faces with light makeup are better recognized than faces with heavy makeup. *Frontiers in Psychology, 7*, 1–8. https://doi.org/10.3389/fpsyg.2016.00226

Chapter 6
Behavioral and Emotional Dimensions of Sexual Harassment at Workplace

Seema Sahai, Richa Goel, and Mahima Punni

Contents

6.1 Introduction

Something that sounds terrible and the first thing that comes to our mind is that why would anyone even do that? With the level of education, values that we have inculcated from our parents and teachers make us question how can anyone have the urge to harass someone sexually? Does it really happen? How often does this happen?

Sexual harassment has been the recent talk of the town and has been the major goal for various companies for 2019. With mandatory rules, training to the employees, etc. the awareness about the same, somewhat the scenario has been improving but at a very slow rate. It can be termed as any physical contact, demand or request, sexual remarks, display of pornography, or any other physical, verbal or nonverbal conduct of sexual nature that is unwelcoming.

There are two types of sexual harassment at the workplace:

1. Quid pro quo—This relates to the exchanging of sexual services for gain such as promotions, rewards, etc. or a loss such as demotion, loss of a job, loss of opportunity, etc.

S. Sahai (✉) · R. Goel · M. Punni
Amity International Business School, Amity University Uttar Pradesh, Noida, India
e-mail: ssahai@amity.edu; rgoel@amity.edu

S. N. Mohanty (ed.), *Emotion and Information Processing*,
https://doi.org/10.1007/978-3-030-48849-9_6

2. Hostile work environment—this relates to an uncomfortable or intimidating environment around the victim, does not always have to be sexual nature, but may also include offensive comments about their sexuality or stereotyping.

Various laws in India related to sexual harassment at workplace—Sexual harassment at the workplace was earlier not conceived as a separate offense under the Indian Penal Court. The first light on the matter was thrown back in 1997, when an NGO, called Vishakha took a gang rape case in Rajasthan to the supreme court. There were other cases too like Mrs. RupanDeol Bajaj v. Kanwar Pal Singh Gill which made the lawmakers feel that there was a need for a separate law dealing with sexual harassment at workplace.

After 16 years, The Sexual Harassment of Women at Workplace (Prevention, Prohibition and Redressal) Act, 2013 came into existence. The law states that physical contact between the victim and the aggressor is not required for it to be termed as sexual harassment. Anything that creates a hostile environment for the victim be it verbal abuses, jokes gestures or rumors to tarnish the victim's image all fall under the scope of the act. A person found guilty would be imprisoned for 3 years under IPC section 354 (modified), in rape cases, maximum punishment or life imprisonment is decided under section 376 except when the victim dies or is left in the inhumane state. Any matter be it small or large needs to be addressed immediately by the management, in countries like Australia the matter falls under Equal Opportunity Act and is strictly compiled with covering the volunteered and unpaid workers as well. Some matters may also be of criminal nature like obscene communication, letters, and threats. These things also effect the state of mind of the victim (Mohanty & Suar, 2014). The mood of the aggressor is also an important issue and has to belooked into and how it gets influenced by environment (Mohanty, Pratihar, & Suar 2015).

6.1.1 Start of the MeToo Movement

MeToo movement was started against Harvey Weinstein, a Hollywood producer who was accused of sexually misbehaving including raping over 70 women, soon after many other women started to raise their voices. It reached India a year after with actresses like Tanushree Dutta and Kangana Ranaut all over the media with their stories. This movement dragged famous celebrities, actors, directors, politicians, etc. like Sajid Khan, Alok Nath, etc. to the limelight for their obnoxious behavior at workplace.

This was high time that sexual harassment had to be addressed (Figs. 6.1, 6.2, and 6.3).

India BSE 100 companies reported 14% rise in sexual harassment complaint cases in the financial year 2019, showing the growing awareness but also requires speedy action against the aggressor.

There were about 823 cases reported in FY19 against 722 in FY18. Wipro reported the highest number of complaints, ICICI and Infosys showed a reduction in the same (Figs. 6.4 and 6.5).

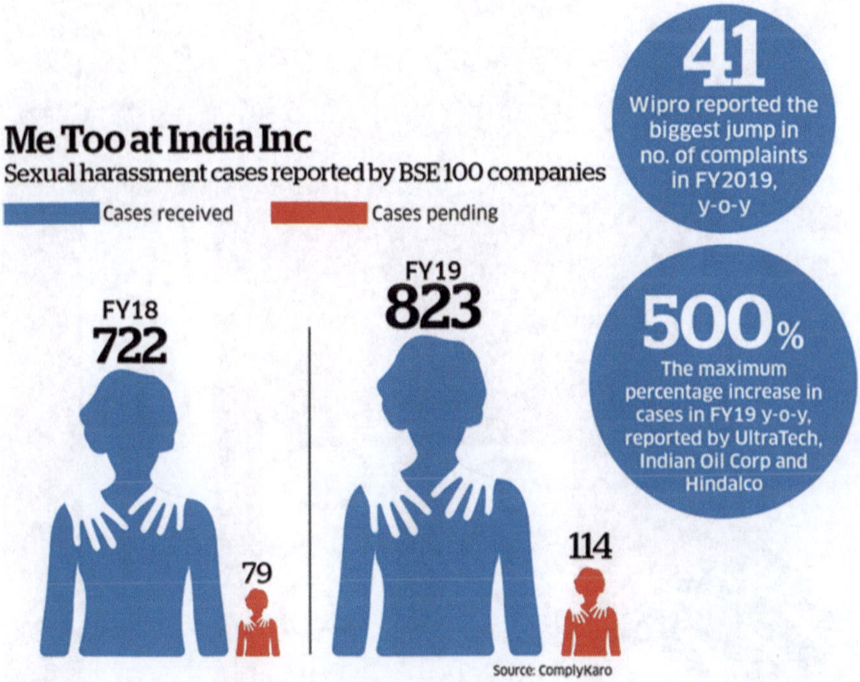

Fig. 6.1 Sexual Harassment case. (Source: Complykaro)

6.2 Review of Literature

1. ArchitaMohapatra, Ajay Solanki and Vikram Shroff. (2019). India: Increased complaints of sexual harassment: Increased number of reported complaints are probably the result of various movements like the #Metoo movement. The act of 2013 makes it mandatory for the employers to frame policies prohibiting sexual harassment at the organization.
2. Society for Human Resource Management. (2018). Harassment-Free Workplace Series. A focus on sexual harassment: with laws strengthening and focusing more on sexual harassment at the workplace, HR professionals of the organizations believe that they know what is going on, whereas as the majority of the incidents go unreported as people are either too scared or unaware of the policies.
3. Melissa Jun Rowley. (2018). How to hold employers accountable during the #TimesUp and #MeToo movement. Forbes: As the powerful message by Oprah Winfrey "a new day is on the horizon" for women became popular and the Time's Up gain traction, workplaces have started responding to domestic and sexual violence and dealing with it at its core by inducing in the company culture. Firing the harasser does not address the organization's systematic problem.

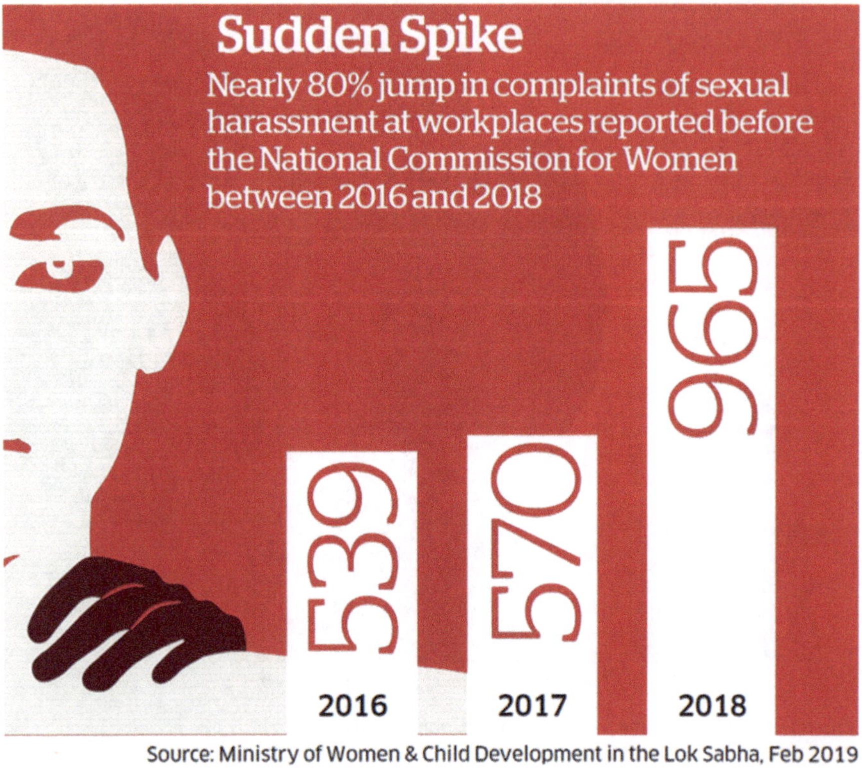

Fig. 6.2 Sudden Spike in reported cases. (Source: Ministry of Women & Child Development in the Lok Sabha, Feb 2019)

4. Bryce Covert (2018). When harassment is the price of a job. The Nation: 60% of the women working at bars and restaurants have said they have faced sexual harassment and that the culture and income are tied to it, the jobs being easy to secure.
5. MaadhuBalaaji S and S. Malaimagal (2018). A study on sexual harassment on women in India: Women across the globe have tasted all flavors from respect and glory to these phases where they are raped, abused, molested, etc. An emphasis has been laid on the psychological, emotional, and mental impacts women suffer because of the sexual abuse.
6. Kundu, Bansal, and Jambheshwar (2018). A critical analysis of sexual harassment of women in India. Researchgate.net: laws, issues, and preventative measures: The study focuses on how sexual harassment affects the dignity of women and the measures to safeguard the same. It indicates that the issue is quite alarming and needs to be carefully attended.
7. Higgs, Joahanna and Jeffries, Brittany. (2018). Sexual Harassment in India. *International Journal for Intersectional Feminist Studies*, Volume 4, Issue 1 & 2, ISSN 2463-2945: Sexual harassment is a problem that is all pervasive throughout the world, the study draws attention to the awareness and the level

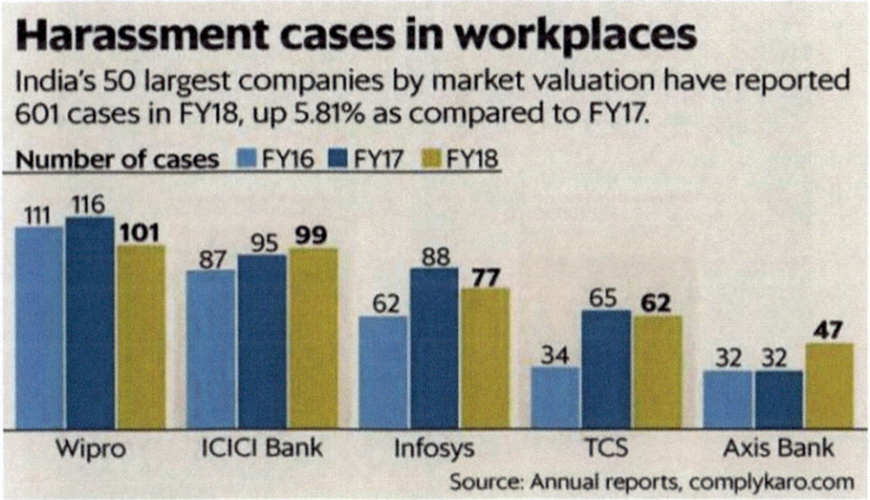

Fig. 6.3 Number of cases from 2016 to 2018. (Source: Annual reports, complykaro.com)

of misconduct being faced by the women in India and turns out there is a long to end it in the country.

8. Susan Chiran and Catrin Einhorn. (2017). How tough is it to change a culture of harassment? Ask women at Ford. *New York Times*: women at Ford were treated like property or prey by bosses and fellow laborers, comments on breasts and buttocks, graffiti of penises, etc. were seen. Supervisors traded rewards for sex and punishment for everyone who refused.
9. Rebecca Traistor. (2017). This moment is not just about sex. It is really about work. The Cut: It is just not about sexual crimes, but it is about women equality at the work place, conditions that make it hard for women to work.
10. Frank Dobbin and Alexandra Kalev. (2017). Training programs and reporting systems will not end sexual harassment but promoting more women will. *Harvard Business Review*: We already know strengthening policies will reduce sexual harassment, but hiring and promoting more women will address it at the grassroots level.
11. Sachi Nandan Mohanty and Damodar Suar (2013a, b). Negative affectivity of participants influence risk seeking on about half of the hypothetical situations. This is the reason why one has to be careful while jumping to conclusions.

6.3 Objectives

The purpose of this study is to analyze and see the impacts of various factors affecting sexual harassment at the workplace. The study starts with understanding what is sexual harassment, current laws, and statistics. Now the study would try to understand what people actually feel would affect the scenario of sexual abuse at the

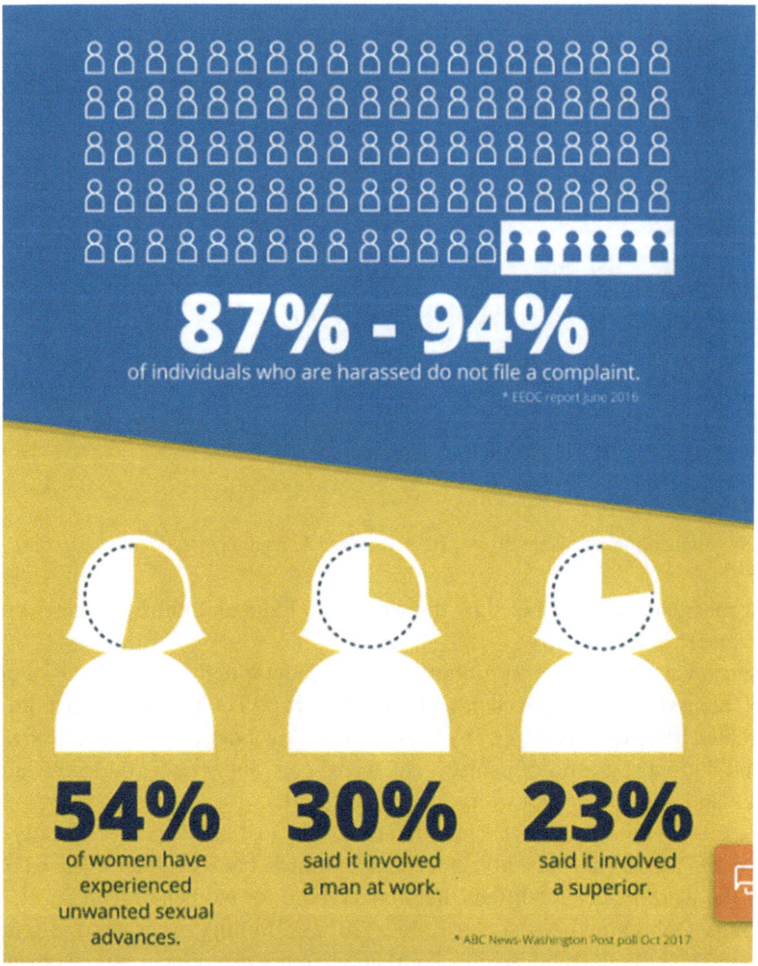

Fig. 6.4 People who do not file a complaint

workplace and what would have no significant impact. The objectives can be defined as:

1. Whether gender plays a role in sexual harassment at the workplace?
2. Whether the level of literacy plays an important role in sexual harassment at the workplace?
3. Whether better management play an important role in sexual harassment at the workplace?
4. Whether punishment would affect sexual harassment at the workplace?
5. Whether awareness about sexual harassment at the workplace would help?
6. Should companies invest in training employees to deal with sexual harassment at the workplace?

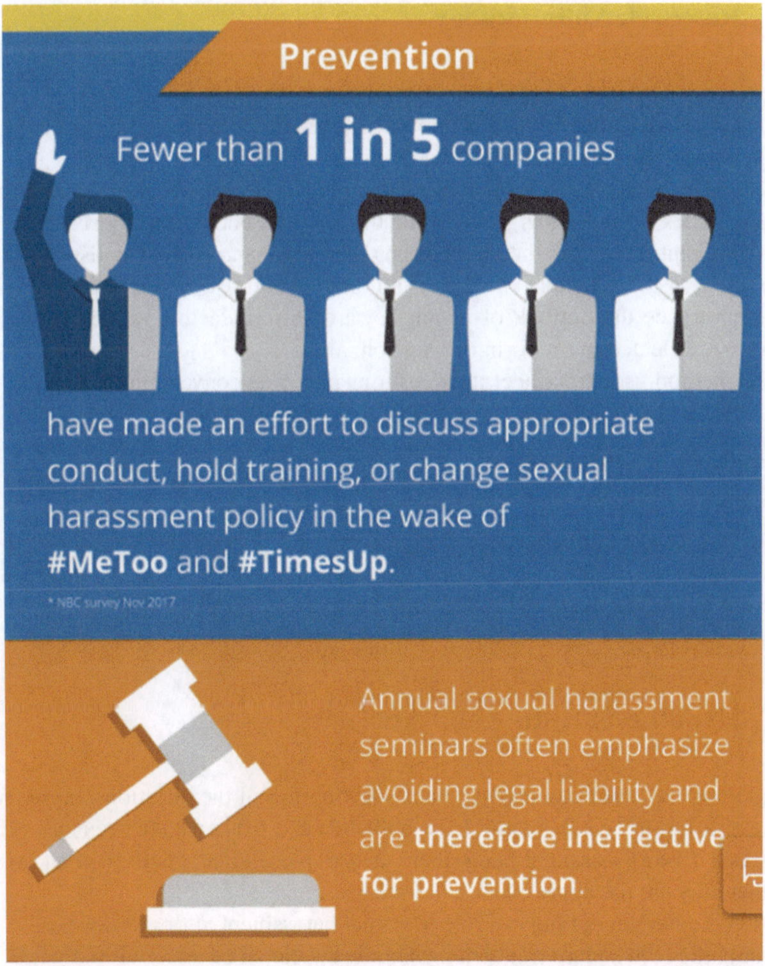

Fig. 6.5 Companies creating awareness

7. Does having a separate properly functioning cell with an open door policy would help in dealing with sexual harassment at the workplace?
8. Whether position and authority play a role in sexual harassment at the workplace?
9. Do employees feel comfortable about standing against sexual harassment at the workplace?
10. Would employees raise their voice if safety and security would be ensured by the organization?

6.4 Research Methodology

There are usually two methods to collect data:

Primary Data Collection—where first-hand data is obtained from various sources like personal interviews and questionnaires.

Secondary Data Collection—where data which is already available is used for further studies from various sources like the internet, books, and journals.

A combination of both the methods has been used, and current statistics of sexual harassment in India have been taken from internet and research reports and respondent data to see the outlook of people toward various factors related to the same have been collected in the form of questionnaires.

The method used to collect the questionnaires was convenience sampling, Data was collected from the nearest mall as well as closest locality.

- Pacific Mall, Subhash Nagar
- Janakpuri D-Block, New Delhi

This is also a combination of exploratory as well as descriptive research, as it has the elements of exploring the issue of sexual harassment and people's outlook toward it as well as describing the same. Sample size: 100 respondents.

6.5 Hypothesis Testing and Analysis

Based upon the questionnaire hypothesis testing for all the questions answered has been done, it is to be kept in mind that both men and women have been taken for the study in 1:1 ratio, as sexual harassment at workplace affects both men and women whether they are the aggressors or the victims.

Q1. Whether gender plays a role in sexual harassment at the work place?

Doing the "chi square test" with a significance level of −5%

H_0= The people saying yes are correct
H_A = The people saying yes are not correct

The degree of freedom is at 5% significant level value is 3.841, which is greater than the calculated X^2 value 0.542, which implies that we would accept the null hypothesis that states that the people saying gender plays a huge role in sexual harassment at work place are correct as it has been seen that most number of harassment cases have women victims, males are also there who have been sexually abused but the number is so low that people usually believe that it is not happening, also the fact that male resist reporting the issue due to the fear of image (Tables 6.1 and 6.2).

Q2. Whether level of literacy plays an important role in sexual harassment at the workplace?

Doing the "chi square test" with a significance level of −5%

Table 6.1 Response of the men and women for and against gender playing a role in sexual harassment at the work place

Response	Yes	No	Total
Male	45	5	50
Female	47	3	50
92	8	100	

Table 6.2 Chi square test on the data taken from Table 6.1

Observed	Expected	$(O - E)^2$	$(O - E)^{13}/E$
45	46	1	0.021
47	46	1	0.021
5	4	1	0.25
3	4	1	0.25
			0.542

H_0 = The people saying yes are correct
H_A = The people saying yes are not correct

The degree of freedom is at 5% significant level value is 3.841, which greater than the calculated X^2 value 0.162, which implies that we would accept the null hypothesis that states that the people saying education plays an important role in sexual harassment at the workplace is true, although there is not a very huge differ ence between the people who believe that and the ones who do not since how someone teats another human being is based on the value system and the wisdom of an individual (Tables 6.3 and 6.4).

Q3. Whether better management play an important role in sexual harassment at the workplace?

Doing the "chi square test" with a significance level of –5%

H_0 = The people saying yes are correct
H_A = The people saying yes are not correct

The degree of freedom is at 5% significant level value is 3.841, which is greater than the calculated X^2 value 0.21, which implies that we would accept the null hypothesis that states that the people saying management play an important role in sexual harassment at the workplace are absolutely correct, most of the respondents agree that management plays huge role, the ones who do not agree says it depends upon the individuals own mindset and values (Tables 6.5 and 6.6).

Q4. Whether punishment would affect sexual harassment at the workplace?

Doing the "chi square test" with a significance level of –5%

H_0 = The people saying yes are correct
H_A = The people saying yes are not correct

The degree of freedom is at 5% significant level value is 3.841, which is greater than the calculated X^2 value 1.01, which implies that we would accept the null

Table 6.3 Response of the men and women for and against literacy playing a role in sexual harassment at the work place

Response	Yes	No	Total
Male	30	20	50
Female	28	22	50
	58	42	100

Table 6.4 Chi square test on the data taken from Table 6.3

Observed	Expected	$(O - E)^2$	$(O - E)^{13}/E$
30	29	1	0.034
28	29	1	0.034
20	21	1	0.047
22	21	1	0.047
			0.162

Table 6.5 Response of the men and women for and against management playing a role in sexual harassment at the work place

Response	Yes	No	Total
Male	48	2	50
Female	47	3	50
	95	5	100

Table 6.6 Chi square test on the data taken from Table 6.5

Observed	Expected	$(O - E)^2$	$(O - E)^{13}/E$
48	47.5	0.25	0.005
47	47.5	0.25	0.005
2	2.5	0.25	0.1
3	2.5	0.25	0.1
			0.21

hypothesis that states that the people saying punishment would affect sexual harassment at the workplace are correct, the ones do not agree believe that some people will do the crime even if they would get punished for the same (Tables 6.7 and 6.8).

Q5. Whether awareness about sexual harassment at the workplace would help?

Doing the "chi square test" with a significance level of –5%

H_0 = The people saying yes are correct
H_A = The people saying yes are not correct

The degree of freedom is at 5% significant level value is 3.841, which is greater than the calculated X^2 value 0.152, which implies that we would accept the null hypothesis that states that the people saying awareness about sexual harassment at the workplace would help in reducing the same (Tables 6.9 and 6.10).

Table 6.7 Response of the men and women for and against punishment affecting sexual harassment at the work place

Response	Yes	No	Total
Male	50	0	50
Female	49	1	50
	99	1	100

Table 6.8 Chi square test on the data taken from Table 6.7

Observed	Expected	$(O - E)^2$	$(O - E)^{13}/E$
50	49.5	0.25	0.005
49	49.5	0.25	0.005
0	0.5	0.25	0.5
1	0.5	0.25	0.5
			1.01

Table 6.9 Response of the men and women for and against creation of awareness about sexual harassment at the work place to reduce the same

Response	Yes	No	Total
Male	46	4	50
Female	47	3	50
	93	7	100

Table 6.10 Chi square test on the data taken from Table 6.9

Observed	Expected	$(O - E)^2$	$(O - E)^{13}/E$
46	46.5	0.25	0.005
47	46.5	0.25	0.005
4	3.5	0.25	0.071
3	3.5	0.25	0.071
			0.152

Q6. Should organizations invest in training employees to deal with sexual harassment at the workplace?

Doing the "chi square test" with a significance level of −5%

H_0 = The people saying yes are correct
H_A = The people saying yes are not correct

The degree of freedom is at 5% significant level value is 3.841, which is greater than the calculated X^2 value 0.542, which implies that we would accept the null hypothesis that states that the people saying investment in training employees to deal with sexual harassment at the workplace would be extremely beneficial (Tables 6.11 and 6.12).

Table 6.11 Response of the men and women for and against investment on training by the organizations to deal with sexual harassment at the work place

Response	Yes	No	Total
Male	45	5	50
Female	47	3	50
	92	8	100

Table 6.12 Chi square test on the data taken from Table 6.11

Observed	Expected	$(O - E)^2$	$(O - E)^{13}/E$
45	46	1	0.021
47	46	1	0.021
5	4	1	0.25
3	4	1	0.25
			0.542

Table 6.13 Response of the men and women for and against having a separate cell having an open door policy at the organization to deal with sexual harassment at the work place

Response	Yes	No	Total
Male	50	0	50
Female	50	0	50
	100	0	100

Table 6.14 Chi square test on the data taken from Table 6.13

Observed	Expected	$(O - E)^2$	$(O - E)^{13}/E$
50	50	0	0
50	50	0	0
0	0	0	0
0	0	0	0
			0

Q7. Does having a separate properly functioning cell with an open door policy would help in dealing with sexual harassment at the workplace?

Doing the "chi square test" with a significance level of −5%

H_0 = The people saying yes are correct
H_A = The people saying yes are not correct

The degree of freedom is at 5% significant level value is 3.841, which is greater than the calculated X^2 value 0, which implies that we would accept the null hypothesis that states that the 100% respondents think that having a separate properly functioning cell with an open door policy would help in dealing with sexual harassment at the workplace, as more and more people would feel comfortable in sharing their experiences without the thought of being judged as the cell is separately made to address these concerns (Tables 6.13 and 6.14).

Table 6.15 Response of the men and women for and against position and authority playing a role in sexual harassment at the work place

Response	Yes	No	Total
Male	48	2	50
Female	48	2	50
	98	4	100

Table 6.16 Chi square test on the data taken from Table 6.15

Observed	Expected	$(O-E)^2$	$(O-E)^{13}/E$
48	49	1	0.020
48	49	1	0.020
2	2	0	0
2	2	0	0
			0.040

Q8. Whether position and authority play a role in sexual harassment at the workplace?

Doing the "chi square test" with a significance level of −5%

H_0 = The people saying yes are correct
H_A − The people saying yes are not correct

The degree of freedom is at 5% significant level value is 3.841, which is greater than the calculated X^2 value 0.040, which implies that we would accept the null hypothesis that states that the people saying that position and authority play a role in sexual harassment at the workplace, as women at higher level are less likely to be harassed and usually the women or even men who get harassed are by people who are their seniors (Tables 6.15 and 6.16).

Q9. Do employees feel comfortable about standing against sexual harassment at the workplace?

Doing the "chi square test" with a significance level of −5%

H_0 = The people saying yes are correct
H_A = The people saying yes are not correct

The degree of freedom is at 5% significant level value is 3.841, which is lesser than the calculated X^2 value 6.245, which implies that we would reject the null hypothesis that states that the people saying employees feel comfortable about standing against sexual harassment at the workplace are wrong, people feel really redundant in speaking up if something like this happens to them, it can also be seen by the results that men are less likely to speak up than women (Tables 6.17 and 6.18).

Q10. Would employees raise their voice if safety and security would be ensured by the organization?

Doing the "chi square test" with a significance level of −5%

Table 6.17 Response of the men and women for and against the comfort that the employees feel to stand up against sexual harassment at the work place

Response	Yes	No	Total
Male	5	45	50
Female	15	35	50
	20	80	100

Table 6.18 Chi square test on the data taken from Table 6.17

Observed	Expected	$(O-E)^2$	$(O-E)^{13}/E$
5	10	25	2.5
15	10	25	2.5
45	40	25	0.62
35	40	25	0.62
			6.245

Table 6.19 Response of the men and women for and against assurance of the safety and security of the employees by the organization if they decide to stand up against sexual harassment at the work place

Response	Yes	No	Total
Male	48	2	50
Female	49	1	50
	97	3	100

Table 6.20 Chi square test on the data taken from Table 6.19

Observed	Expected	$(O-E)^2$	$(O-E)^{13}/E$
48	48.5	0.25	0.005
49	48.5	0.25	0.005
2	1.5	0.25	0.16
1	1.5	0.25	0.16
			0.336

H_0 = The people saying yes are correct
H_A = The people saying yes are not correct

The degree of freedom is at 5% significant level value is 3.841, which is greater than the calculated X^2 value 0.336, which implies that we would accept the null hypothesis that states that the people saying employees raise their voice if safety and security would be ensured by the organization are absolutely correct, usually people feel that if they would speak up and there is not proper action, the aggressor might start affecting them outside the workplace as well, things like threatening are expected (Tables 6.19 and 6.20).

6.6 Conclusion

Sexual harassment is very deeply rooted in the country, from rural to urban areas, it is all-pervasive and not just here in India but all across the globe. With the advent of social media, there has been a significant impact in the awareness leading to the growing number of reported claims, but it also shows how the number is still growing and the number of sexual crimes is not going down. Men are shying away from admitting that it is happening with them as well. With more and more power to women, women are being able to speak not a 100% but still show a significant improvement to the earlier times, but the male gender is still not empowered to speak upon this topic (about 75% of individuals who face the issue are reluctant to speak). Women are being placed at higher authority levels which is depicting that the situation is getting better, but sadly there is a long way ahead, and it would require decades to completely eliminate objectifying and molesting women.

Based on the study, it can be seen that if organizations would be ready to ensure safety and security then more and more people would be able to report their issues, also if there would be a separate cell, people would feel more comfortable going to them rather than talking to the HR professionals who are dealing with routine nature human resource management. There is also a need for training employees with various techniques to deal with misconduct like self-defense classes, regular seminars to inform about the policies to deal with such issues along communicating the rights and laws to the employees should be conducted. At last I would just like to conclude by saying organizations can do their part but inducing the right attitude begins at home, parents need to ensure that they teach their children to respect and behave properly, if every parent would monitor their child and would ensure his behavior or attitude is no way alarming of such desires, then the day would not be far where every individual would respect not only their family members but also the people around them, leading to a congenial environment as a whole.

Annexure

QUESTIONNAIRE

PLEASE FILL THE QUESTIONNAIRE BELOW TO HELP ME COLLECT DATA FOR MY COLLEGE REPORT ON SEXUAL HARASSMENT AT WORKPLACE. You are not required to give your name or any other personal details except mentioning your gender.

GENDER__________

1. Whether gender plays a role in Sexual harassment at the workplace?

YES ☐ NO ☐

2. Whether level of literacy plays an important role in Sexual harassment at the workplace?

YES ☐ NO ☐

3. Whether better management play an important role in Sexual harassment at the workplace?

YES ☐ NO ☐

4. Whether punishment would affect Sexual harassment at the workplace?

YES ☐ NO ☐

5. Whether awareness about Sexual harassment at the workplace would help?

YES ☐ NO ☐

6. Should companies invest in training employees to deal with Sexual harassment at the workplace?

YES ☐ NO ☐

7. Does having a separate properly functioning cell with an open door policy would help in dealing with Sexual harassment at the workplace?

YES ☐ NO ☐

8. Whether position and authority play a role in Sexual harassment at the workplace?

YES ☐ NO ☐

9. Do employees feel comfortable about standing against Sexual harassment at the workplace?

YES ☐ NO ☐

10. Would employees raise their voice if safety and security would be ensured by the organization?

YES ☐ NO ☐

References

Covert, B. (2018). *When harassment is the price of a job*. Retrieved from https://www.thenation.com/article/archive/when-harassment-is-the-price-of-a-job/

Kundu, S. C., Bansal, J., & Jambheshwar, G. (2018). *A critical analysis of sexual harassment of women in India: Laws, issues, and preventive measures ability-motivation-opportunity enhancing HR practices in India View project*. Retrieved from https://www.researchgate.net/publication/325712751

Mohanty, S. N., Pratihar, D. K., & Suar, D. (2015). Study on influence of mood states on information processing during decision making using fuzzy reasoning tool and neuro-fuzzy system developed based on Mamdani approach. *International Journal of Fuzzy Computing Modeling, 1*(3), 252–268.

Mohanty, S. N., & Suar, D. (2013a). Decision-making in positive and negative prospects: Influence of certainty and affectivity. *International Journal of Advances in Psychology, 2*(1), 19–28.

Mohanty, S. N., & Suar, D. (2013b). Influence of mood states, group discussion, and interpersonal comparison on change in decision-making and information processing. *International Journal of Decision Sciences, Risk and Management, 5*(2), 101.

Mohanty, S. N., & Suar, D. (2014). Decision making under uncertainty and information processing in positive and negative mood states. *Psychological Reports, 115*(1), 91–105.

Chapter 7
The Effect of Age, Gender, and Arousal Level on Categorizing Human Affective States

Hemanta Kumar Palo

Contents

7.1 Introduction

Robert Plutchik has introduced "wheel of emotions" to classify emotions into eight basic categories in line with six basic emotional categories suggested by Paul Ekman (Plutchik, 2001). The six basic emotional categories are anger, fear, sadness, happiness, disgust, and fear as shown in Fig. 7.1.

Grouping of emotions into positive and negative classes such as trust versus mistrust, anger versus fear, joy versus sadness, and surprise versus anticipation has been made as a step to distinguish different emotions. Other secondary emotions have been a blended form of two or more primary emotions, for example, contempt is derived by mixing anger and disgust emotions. Based on the emotional distance between emotional experiences, further classifications of human emotions into two dimensions have explored in pieces of literature (Mohanty & Palo, 2020; Ram, Palo, & Mohanty, 2013). Two different conditions as valence and activation have been used to demarcate emotions into two sets. Valence indicates the amount of negative or positive feelings one has for a particular emotion, whereas activation

H. K. Palo (✉)
Department of Electronics and Communication Engineering, ITER, Siksha 'O' Anusandhan (Deemed to be University), Bhubaneswar, Odisha, India
e-mail: hemantapalo@soa.ac.in

S. N. Mohanty (ed.), *Emotion and Information Processing*,
https://doi.org/10.1007/978-3-030-48849-9_7

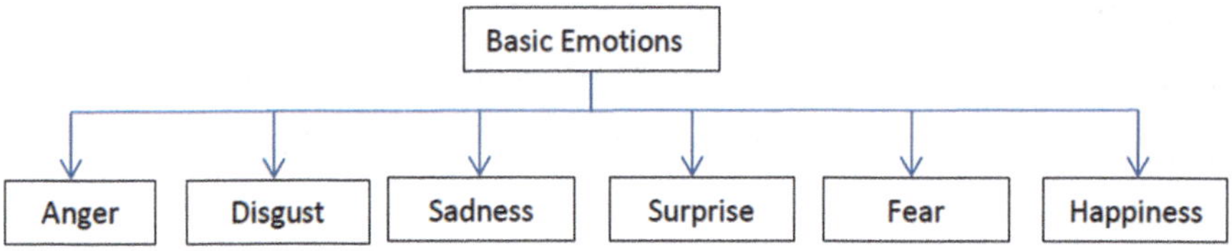

Fig. 7.1 The basic human affective states

relates to the arousal or energy level of a particular category. Emotions such as joy, fear, and anger found to excite the sympathetic nervous system resulting in higher heartbeats, dried mouth, increased blood pressures, larger sub-glottal pressure, and an increase in respiratory functions as suggested in works of literature (Raj & Kumar, 2015). These are higher arousal emotions being fast, higher pitch, energy, and larger pitch range. Low arousal emotions on the other hand excites the parasympathetic nervous system and has the opposite characteristics than the former. These emotions have lower intensity and energy and are slow. Sad and boredom are considered in this category. However, all the high-arousal emotions do not have a higher activation level. For example, the difference between happiness and anger can be distinguished by their valance level (Paulmann, Bleichner, & Kotz, 2013). Thus, it possesses another challenge in demarcating high arousal emotions.

7.2 Application Domain of Speech Emotion Characterization

There have been numerous applications on the recognition of human affective states during a conversation in daily lives. Call center application, medical emergency kiosk, security, the application under abnormal threat conditions, assessment of psychiatric cases, computer games, the interaction of a child with computer characters, tutoring systems, aircraft cockpits, etc. are few such applications where the system can be employed for fruitful results (Bahreini, Nadolski, & Westera, 2016; Mohanty & Palo, 2019; Palo & Behera, 2020; Palo & Mohanty, 2018; Ram et al., 2013). A few major applications are briefed below.

7.2.1 Tutoring System

An effective tutoring system more or less relies on the state of mind of the student during interactive sessions existing with the computers. The emotional content of the student such as positive or negative can be an effective feedback mechanism for the teacher to plan for teaching and evaluation methods and enhance a student's learning potential (Ai et al., 2006; Bahreini et al., 2016). Audio recordings of multimodal interfaces can serve as a yardstick for performance measures during a classroom session for an intelligent teacher. A computer tutorial application requiring natural man–machine interaction has been a major implication for students.

7.2.2 Lie Detection

Criminal psychologists are bound to emphasize on lie detectors in this complex world to study human psychology (Frank & Feeley, 2003). Agencies like police department, security sections, FBI, CBI, RAW, IB, and other intelligent wings will be on an advantage stage, provided the Emotion Recognition System (ERS) is accurate to detect a lie based on human psychology. The recognition system can help investigators finding corruption related to betting in cricket and other sports, money laundering activities, etc. based on the emotional expression during an interaction.

7.2.3 Banking

A speaker authentication/recognition system combined with an emotion recognition facility can be a major boost for ATMs. It can be a boon for banking in delivering confidential information by ensuring an enhanced security level. A machine capable of speaking, understanding, and recognizing human speech with emotions is another major step in this area. To access bank officials using the emotional outputs during the customer handling activities may provide better customer satisfaction in the long run.

7.2.4 Recruitments

It will be wonderful if human resource planning, enrollment of job aspirants without human assistance can be done effectively with the help of ERS. The system can study the behavioral aspects of the job seeker such as nervousness, anxiety, or anger during an interview without any supervisory assistance. On the other hand, it helps to prevent bias against any fraudulent selection due to human favor. Further proxy enrollment can be avoided based on a voice print of the enrollee or client recorded during preliminary stages.

7.2.5 Security

An ERS can be an asset for people in danger (Hernández et al., 2016). A security system can greatly rely on it. By detecting emotions such as fear, stress, or panic of the security personnel deployed at various places, potential situational evidence can be detected and appropriate actions can be taken. In case of hijacking of aircraft, bus, rail, or any transport system, the machine can give considerable information about the situation in hand. Other untoward incidence based on signal systems, malfunctioning of equipment of the transport systems or any potential hazards can be detected beforehand and can be tackled in time.

7.2.6 In-Car Board System

Information on a driver's emotional state can provide input to the in-car board system to actuate safety measures instantly based on situations. It can also facilitate resolving major errors beforehand based on the emotional content of the driver during communication. The application of the in-car-board system to initiate automatic action for the safety of the driver based on certain emotions has been one such attempt in this direction (Tawari & Trivedi, 2010).

7.2.7 Prosody in Dialog System

The prosodic features such as amplitude, frequency, energy, pitch, speech rate, and formants can help to develop a model based on speaking style, language, age, gender, and so on for institutional and industrial use (Palo, Mohanty, & Chandra, 2018; Paulmann et al., 2013). These features can distinguish neutral voices from other emotions, hence can serve teaching and research activities. The recognition system based on discriminating features describing speech emotions provides significant inputs in the ticket reservation system. Similarly, the frustration or annoyance associated with the Prosody for dialog systems, anger management, student counseling, and human resource management can be benefited from exploring reliable speech prosody containing human affective states (Palo, Chandra, & Mohanty, 2017; Palo, Kumar, & Mohanty, 2017; Palo, Mohanty, & Chandra, 2016).

7.2.8 Emotion Recognition in Call Center

In today's tough competition, industries rely on call centers for the effective advertisement of their product. They desire the call center executive to be polite, amicable, and pleasing in explaining their product even to an annoyed customer. Loss of business as a result of customer discontent is a common practice today due to inefficient call center executives that lead to major disputes in industries. An ERS can answer to assess the performance of call center employees. The planning for promotions, increments, and performance bonuses of an employee can be made based on an accurate identifier of speech emotions during a call interaction. Planning, organizing, and placing for recruitment and training of employees can be made based on the feedback of the recognition system. For example, changes in short time energy, rate of speech and pitch contour irregularities of customer annoyance or frustration can provide an early warning for major strategic planning in these centers (Petrushin, 2011).

7.2.9 *Voice Mail*

An electronic system associated with the emotion detection system for recording and storing of the human voice will help a person to respond based on the urgency of the caller (Martinez, 2017). Based on emotional content, several untoward situations can be avoided in time, even if the person is unable to attend the call during his absence but have time to respond later on.

7.2.10 *Computer Games/Web Movies*

Today computer games are a major source of entertainment in all ages of life. The level of the game such as hard, medium, or easy can be controlled based on human emotions that the computer recognizes. As an example, the computer shifts the level to hard or easy based on the aggressiveness of the player. Similarly, people interested in designing web movies must take into account viewers' emotions and change the direction of movies according to their expectations (Merkx, Truong, & Neerincx, 2007).

7.2.11 *Diagnostic Tool by Speech Therapists*

People affected by speech, language, or voice disorder can be benefited from the use of ERS. It can be a handy tool for the speech therapist to treat with understanding the type of strain or stress the affected person is going through. The prosodic features providing information on the vocal tract can be an asset to clinical analysis. With the help of a glottal waveform, considerable information on the affected states of the patients can be obtained (Kostoulas et al., 2012).

7.2.12 *Robots*

Today, robots are anywhere and everywhere interacting and assisting people in their workspace, in industries, shops, supermarkets, homes, and hospitals. It is therefore essential for the robots to understand and recognize human emotions for fulfilling the expected task in hand. An effective ERS thus remains a vital tool for these robots during interaction with a human being in a user-friendly atmosphere (Song, Han, & Wang, 2014).

7.3 Feature Extraction

Among the traditional feature extraction techniques, prosodic and spectral features have been widely used to characterize and classify emotions in speech signals by many researchers (Palo, Behera, & Rout, 2020; Palo & Mohanty, 2020; Palo & Sarangi, 2020). The speech prosody extracted in general remains supra-segmental. Nevertheless, the number of emotions and feature dimensions tend to increase the complexity of a recognition system. Thus, the extraction and selection of discriminating and reliable features play a central role in designing a suitable recognition model. A few of the state-of-art feature extraction techniques adopted in earlier literature have been provided in Fig. 7.2.

7.3.1 Spectrum Variance

Frequency domain analysis of a non-stationary signal such as speech can provide useful information on the status of an affected person. The vocal cord and the vocal tract vibrate and resonate at a certain frequency, thus help to characterize a speaker's voice. Further, the shape and size of the speech production system vary among people with different ages, gender, and affective states. This has made the spectrum as one of the potential features in representing and classifying emotions from voice inputs. It can be extracted from a signal $s(n)$ using the Fourier transform (FT) as

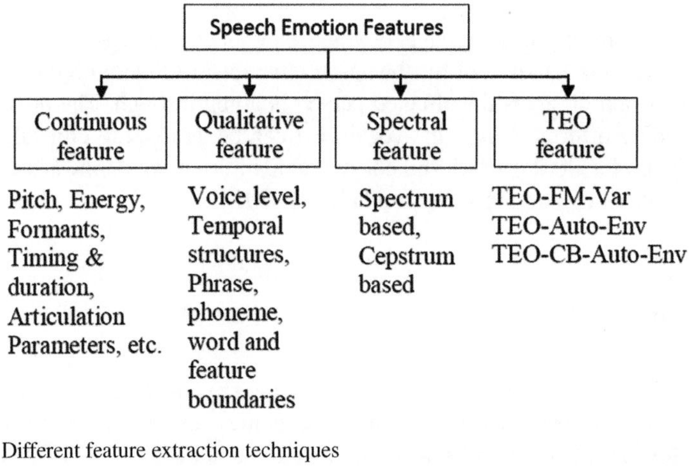

Fig. 7.2 Different feature extraction techniques

$$S(k)=\sum_{n=0}^{N-1}s(n)e^{\frac{-j2\pi}{N}} \tag{7.1}$$

However, the speech signal is aperiodic and can be analyzed effectively over a short frame. This gives rise to the short time FT (STFT) which outperforms the traditional FT in characterizing and classifying a speech sample. The STFT spectrum of an emotional utterance can be estimated from the windowed signal as

$$S(f)=\sum_{n=0} N_a s(t)e^{-j\left(\frac{2\pi f}{f_a}\right)n} \tag{7.2}$$

where f_a and f represent the sampling frequency and the signal frequency of the analyzed signal, respectively. The variable N_a denotes the window duration. Normally, for speech signal, a frame size of 20–50 ms with a frame overlapping of approximately 50% has been found in the literature (Palo et al., 2016). The Hamming window provides the desired smoothness at the signal frame boundary than a rectangular window. It has twice the bandwidth as compared to a rectangular window, hence mostly adopted by speech engineers.

7.3.2 Cepstrum Magnitude

The hearing mechanism of a human being is logarithmic. Thus, approximating the logarithmic human hearing mechanism with compatible features can be more effective and discriminating. This has made the cepstrum techniques as the state-of-the-art approach in the field of ERS. Further, the use of the logarithm concept assists the cepstral analysis to segment the system and source contents of a voice signal without any knowledge on either the system or the source a priori. The voice signal is considered as a convolution of excitation source $s_1(n)$ and vocal tract filter $s_2(n)$ as given by $s(n) = s_1(n) * s_2(n)$. In the frequency domain, it can be represented as

$$S(\omega)=S_1(\omega)\cdot S_2(\omega) \tag{7.3}$$

The logarithmic representation of Eq. (7.3) is

$$\log|S(\omega)|=\log|S_1(\omega)|\cdot\log|S_2(\omega)| \tag{7.4}$$

The IDFT of the log spectrum provides the cepstrum of a signal and is written by

$$\text{IDFT}\left(\log|S(\omega)|\right)=\text{IDFT}\left(\log|S_1(\omega)|\right)+\text{IDFT}\left(\log|S_2(\omega)|\right) \tag{7.5}$$

7.3.3 Autocorrelation Coefficients (ACF)

The auto-correlation analysis provides information on the periodicity of a voice sample. It can be estimated with a time lag τ as

$$A(\tau)=\frac{1}{N}\sum_{n=0}^{N-1}s(n)s(n+\tau) \tag{7.6}$$

The use of ACF in characterizing a signal colored with affective states can provide the correlation among these states. The ACF provides the information on the energy of a signal, hence varies based on the arousal level of emotion.

7.3.4 Pitch or Fundamental Frequency (F0)

The pitch varies between gender, speakers, affective states, and the age of a speaker. The ACF method of extracting pitch has been adopted here due to hardware simplicity, straightforwardness, reliability, phase insensitivity, and robustness. Further, the computation of the ACFs directly from the signal waveform and the use of a single multiplier with an accumulator makes it simpler than other techniques. The ACF attains the highest value when $s(n) = s(n + \tau)$ and the peaks occur at $\tau = IT$, where T is the period and I is an integer. The F0 is calculated from the peak location at $\tau = T$.

$$\text{Pitch period}(m)=\arg_m \max A_m(\tau) \tag{7.7}$$

where $m = 1, 2, \ldots, M$ denotes the length of the analyzed signal. The fundamental frequency can be computed as

$$F_0=\frac{\text{Signal frequency}}{\text{Pitch period}} \tag{7.8}$$

7.3.5 Speech Rate

The rate at which human being speaks varies among the affective states, hence can provide important emotional cues. It provides the time to communicate at the instant of conversation. It represents the number of words or syllables or spoken units uttered per second or minute by an affected person. It indicates the speed or quickness during the utterance of a sentence when the speaker is under different moods.

It is an utterance-level feature. The average speaking rate of a speech sample is computed as

$$R(s) = \frac{N_V(s)}{D(u)} \tag{7.9}$$

Here $N_V(s)$ denotes the vowel segment and $D(u)$ is the duration of a speech signal.

7.3.6 Zero-Crossing Rate (ZCR)

ZCR indicates the rapid fluctuation as the signal makes the transition between positive and negative around zero. It can be represented as

$$\mathrm{ZCR}(n) = \sum_{n=-\infty}^{\infty} \left| sgn(s(n)) - sgn(s(n-1)) \right| \tag{7.10}$$

In general, the fluctuation around zero axes remains more with the high-energy or high-frequency signal. Thus, the level of arousal of emotion can be easily distinguished by studying the ZCR.

7.3.7 Log Energy

The energy of a signal indicates the arousal level of a human affective state. The feature indicates the volume and loudness of the affected speech sample. Further, the feature can approximate the human hearing mechanism as it is logarithmic. Hence, the log energy of a signal provides valuable information on an emotion. It is computed as

$$e_{\log} = 10 \log \sum_{n=1}^{\infty} \left| s(n) \right|^2 \tag{7.11}$$

However, a speech sample can be better described over short segments. Thus, the Short Time Energy (STE) is extracted from each frame of a signal by applying a suitable windowing. The STE of a windowed signal with analyzing window $w(m)$ can be computed as

$$e_{\mathrm{ST}}(m) = \sum_{n=-\infty}^{\infty} \left| s(n) w(m-n) \right|^2 \tag{7.12}$$

7.3.8 Spectral Centroid (SC)

The SC parameters are computed by segmenting the voice sample into some fixed number of sub-bands. The number of sub-bands, the shape of the sub-band filter, and the corresponding center cut-off frequencies play a major role in the computation of SC. It can be calculated using the following steps.

- Initially compute the power spectrum of the voice sample
- Input the computed power spectrum to the designated filter banks
- Find the sub-band centroid or the first moment

From the short time Fourier transform, power spectrum $S_F[n]$ is computed to find the SC feature as

$$\mathrm{SC}_F = \frac{\sum_{n=1}^{N} S_F[n] * n}{\sum_{n=1}^{N} S_F[n]} \tag{7.13}$$

where F is the number of frames of a signal. The SC provides a balancing point of the spectral magnitude. It can capture the information on the tilt or spectral slope similar to the SE within a frame.

7.3.9 Spectral Flux (SF)

The change in local spectral characteristics of a speech sample can be observed with the SF features. It is computed as

$$\mathrm{SF}_F = \sum_{n=1}^{N} \left(S_F[n] - S_{F-1}[n] \right) \tag{7.14}$$

where $S_F[n]$ and $S_{F-1}[n]$ represent a signal spectrum normalized magnitude at two adjacent frames, respectively.

7.3.10 Spectral Roll-Off (SR)

The frequency below which most of the total energy is contained can be expressed as a fraction and is termed as spectral roll-off. The range of the fraction or threshold is considered to be within 0.85–0.95 as per the requirement or the situation in hand. Incidentally, the occurrence of the more high-frequency component in an affective state is governed by the shape of the spectral. Thus, the SR can provide spectral shape properties. It can be computed as

$$\sum_{n=1}^{K_t} S_F[n] = 0.85 * \sum_{n=1}^{N} S_F[n] \tag{7.15}$$

where $S_F[n]$ represents the frequency below which 0.85 fractions of $S_F[n]$ resides.

7.3.11 Skewness

For a speech signal, the skewness indicates the symmetry of the distribution of the features around a certain point. If the features are normally distributed or symmetrically distributed, the skewness is considered to be near zero. If the feature values are negative, it indicates the data are left-skewed (with long left tail) and vice versa. It can be calculated using the Fisher–Pearson coefficient as

$$\text{Sk} = \sum_{j=1}^{M} \frac{\left(s_j - \overline{s}\right) / M}{d^3} \tag{16}$$

where d, $\overline{s}$, and M denote the standard deviation, mean, and the number of features extracted from a speech signal s.

7.3.12 Kurtosis

Whether the features representing an emotional speech sample are light or heavy-tailed corresponding to the normal distribution is indicated by kurtosis. A larger kurtosis indicates that the features have outliers or heavy tails and vice versa. In an extreme case, the features may be uniformly distributed. The kurtosis can be computed as

$$\text{kt} = \sum_{j=1}^{M} \frac{\left(s_j - \overline{s}\right)^4 / M}{d^4} \tag{7.17}$$

In the case, the features are normally distributed, the kurtosis remains zero.

7.3.13 Shannon Entropy

The Shannon entropy of a signal can be computed as

$$H(\text{Shannon}) = \sum_j p_j \log_b p_j \tag{7.18}$$

where p_j denotes the probability of a feature value occurring in the feature set extracted from an emotional speech sample.

7.3.14 Features of the Vocal Tract Model

When excited by the sound source, the vocal tract cavity acts as a time-varying filter, which resonates at different frequencies due to vocal folds that act as an oscillator. Thus, the vocal tract can be hypothesized as a tube with variable diameters. Accordingly, speech samples can be modeled as the outcomes of a digital filter that is being excited by voicing sounds (equally spaced pulses) or non-voiced (white noise). This is shown in Fig. 7.3.

Speech signals involving nasals and fricative can be represented by a model having both poles and zeros given by

$$H(z) = \frac{S(z)}{W(z)} = G\frac{1+\sum_{l=1}^{r} b_l z^{-l}}{1+\sum_{k=1}^{q} a_k z^{-k}} \tag{7.19}$$

where $S(z) = \sum_{\infty}^{n=-\infty} s_n z^{-n}$ is the output of any system with an unknown input $W(z)$ having gain G with filter coefficients a_k, $1 \leq k \leq q$ and b_l, $1 \leq l \leq r$ 1 with $b_0 = 1$. Nevertheless, the simplified autoregressive (AR) all-pole model can ideally represent the non-nasal voiced sound. However, if the filter of sufficiently higher order is chosen, all speech sounds can be approximated to an all-pole (AR) model given by

$$H_{\text{AR}}(z) = \frac{G}{1+\sum_{k=1}^{q} a_k z^{-k}} = \frac{G}{A(z)} \tag{7.20}$$

This model assumes that any signal s_n can be described by a linear combination of past few samples and the given input with the predictor coefficient a_k. Here,

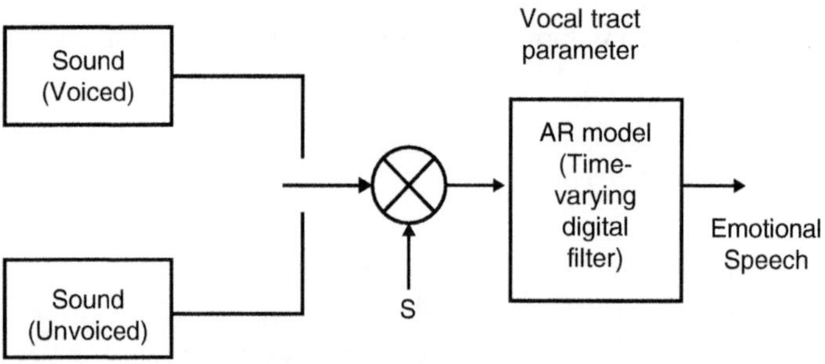

Fig. 7.3 Speech model based on linear prediction

$A(z) = 1 + \sum_{l}^{i=1} a_i z^{-i}$ is generally called as the inverse filter and is identical to the acoustic tube model of the human being.

The spectrum $\hat{S}(\omega)$ of this acoustic model can be expressed as

$$\hat{S}(\omega) = \left|\hat{H}_{\text{AR}}\left(e^{j\omega}\right)\right|^2 = \frac{G^2}{\left|A\left(e^{j\omega}\right)\right|^2} = \frac{G^2}{1 + \sum_{k=1}^{q} a_k z^{-jk\omega}} \tag{7.21}$$

The measurement of error e can be found form the difference between the given speech spectrum $S(\omega)$ and $\hat{S}(\omega)$. This e indicates the total energy of the error signal resulted due to the passing of the speech signal s_n through the filter $A(z)$. Using Parseval's theorem, it can be described by

$$e = \frac{s^2}{2\pi} \int_{\pi}^{-\pi} \frac{S(\omega)}{\hat{S}(\omega)} d\omega = \frac{1}{2\pi} \int_{-\pi}^{\pi} S(\omega) \left|A\left(e^{j\omega}\right)\right|^2 d\omega \tag{7.22}$$

To determine the filter coefficients a_k, the error can be minimized with respect to each filter coefficient. Since e is independent of the gain factor, the minimization can be formalized by setting $\frac{\partial e}{\partial u_i} = 0, 1 \leq i \leq q$. From Eqs. (7.20) to (7.22), we can derive the Eq. (7.23) as

$$\frac{\partial e}{\partial a_i} = 2\left[R_i + \sum_{i=1}^{q} a_k R_{|i-k|}\right] \tag{7.23}$$

Here, R_k is the autocorrelation function is given by $R_k = \frac{1}{2\pi} \int_{\pi}^{-\pi} S(\omega)\cos(k\omega)d\omega$ with regard to speech signal spectrum $S(\omega)$. Equating Eq. (7.23) to zero, we can find the set of q linear equations for q unknowns given by

$$\sum_{q}^{i=1} a_k R_{|i-k|} = -R_i \tag{7.24}$$

This equation can be solved recursively to find the linear predictor filter coefficients. Similarly, the AR model spectrum autocorrelation function can

$$\hat{R}_i = \frac{1}{2\pi} \int_{\pi}^{-\pi} \hat{S}(\omega)\cos(i\omega)d\omega \tag{7.25}$$

The filter-bank approach or LP analysis used to derive cepstral coefficients are treated as standard features for long. Phonetic information is directly extracted from speech spectra, so better described by spectral features than prosodic features [8].

Its extract the feature of the signal. . These features tend to be robust due to the accurate characterization of emotion-specific speech information.

7.3.14.1 Mel-Frequency Cepstral Coefficient (MFCC)

MFCCs have been most popular and robust in recognition of speech signals. The technique approximates the signal linearly at a lower frequency below 1 kHz and logarithmically at higher frequencies using a Mel-scale

$$f_1 = 2595 \log 10\left(1 + \frac{f}{700}\right) \tag{7.26}$$

where f denotes the signal frequency. In the speech signal, the majority of phoneme-related information resides at lower frequencies. The nonlinear Mel-scale tends to emphasize these lower frequency components to extract the relevant information of an utterance. In this, the cepstrum at the Mel-frequency signifies the short-term power spectrum obtained from each speech frame. The MFCCs have been chosen since it takes into account the perception of the nonlinear auditory system of the listener during recognition of speech emotion.

7.3.14.2 Linear Predictor Cepstral Coefficient (LPCC)

LPCCs can be derived from taking the logarithm of LPCs. The p number of LPCC coefficients c_p can be obtained from the q number of LP coefficients a_q using the relation as is given by

$$a_p = \left\{ \begin{array}{ll} a_m + \sum_{p-1}^{k=1}\left(\frac{k}{p}\right) c_k a_{p-k}, & \text{for } 1 \le k \le q \\ \sum_{p-1}^{k=p-q}\left(\frac{k}{p}\right) c_k a_{p-k}, & p > q \end{array} \right. \tag{7.27}$$

LPCCs are more robust than LPCs. The overlapped and correlated features are additive in LPCCs due to the use of logarithms in their extraction. Contrarily, LPCs use a linear scale; hence, these features are multiplicative in modeling the emotions. Thus, LPCCs can provide additional information on two distinct features. The technique helps in separating the vocal tract and excitation source characteristics in the cepstral domain. Further, in this technique, the energy of filter banks is used to extract the features from the signal spectra considering the equal contribution due to each frequency component. This chapter uses a tenth-order LP analysis to extract 16 LPCCs from each frame of an emotional utterance.

7.3.14.3 Perceptual Linear Predictor (PLP) Coefficient

PLP technique differs from the MFCC technique in terms of the frequency scale used and an advanced version of LP analysis. It uses a nonlinear Bark scale $f_2(\omega)$

$$f_2(\omega) = 6\ln\left[\frac{\omega}{1200\pi} + \left[\left(\frac{\omega}{1200\pi}\right)^2 + 1\right]^{1/2}\right] \quad (7.28)$$

It provides the critical band spectral resolution. It also accounts for the equal loudness level and uses the intensity-loudness power law in describing the intended signal. While LP analysis emphasizes only higher frequencies, PLPs deemphasize higher frequencies (F3 and above) and give due importance to the first two formants (F1 and F2). PLP features are highly correlated in nature and their peaks are less sensitive to vocal tract length. Thus, these can distinguish between unvoiced and voiced parts of a signal.

7.3.14.4 Formants

Formants describe the dimension and shape of an individual's vocal tract system. Due to the level of excitation concerning different affective states, there is a change in the vocal tract parameter. Hence, the formants can provide valuable information about different affective states. In general, the first three formants contain most of the speech information; hence, they are considered by researchers in analyzing voice attributes. Formants are confined to the higher amplitude of signal spectra hence not masked by noise. While cepstrum-based features represent the amplitude or energy information of the signal power spectrum, formants use the frequency information.

For a q-order, AR vocal tract model given by $\hat{\phi}(z) == \frac{1}{1-\sum_{k=1}^{q} a_k z^{-k}}$ having LPCs as a_k, $1 \le k \le q$, formants are the angles of the poles which are away from the origin in the z-plane. These are extracted here on solving the roots of LPCs. For a sampling frequency f_s, the formant frequency $\hat{f}$ can be found using the relation $\hat{f} = \frac{f_s}{2\pi}\beta_0 \text{Hz}$. Assuming a second-order LPC model, the formant frequency can be rewritten as

$$\hat{f} = \frac{f_s}{2\pi}\arccos\left[\cos\beta_0 \frac{\left(p_0^2+1\right)}{2p_0}\right] \quad (7.29)$$

where the complex root pairs can be expressed as $g = p_0 \exp(\pm\beta_0)$.

7.4 Results and Discussion

The spectrum variance is maximized for high arousal affective states such as angry and happy as compared to the low arousal states such as bore and sad signals. The availability of high-frequency components is the possible reason for this trend as shown in Fig. 7.4.

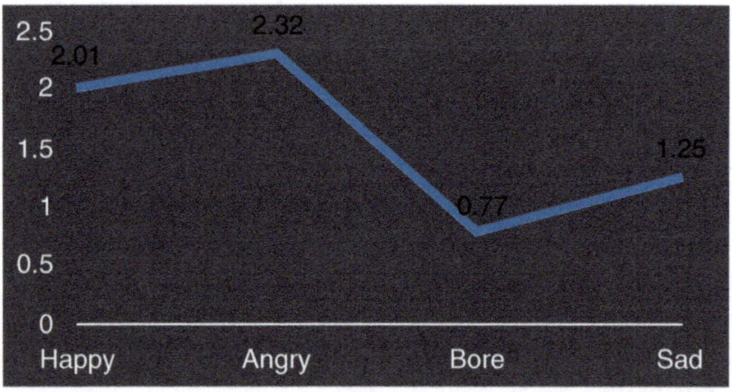

Fig. 7.4 Maximum spectrum variance of different human affective states

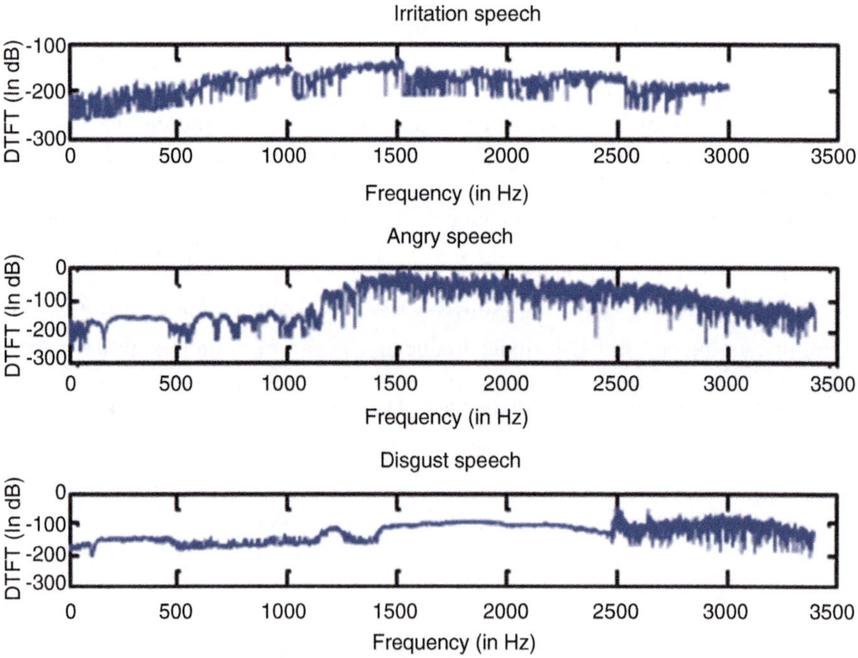

Fig. 7.5 The variation in STFT among different affective states

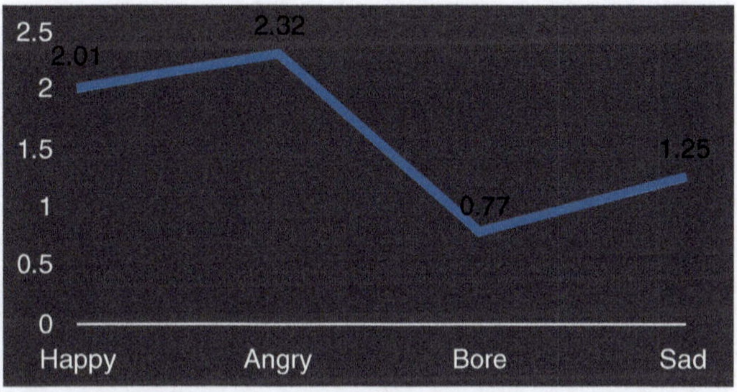

Fig. 7.6 Maximum cepstrum magnitude of different human affective states

The similarity among an angry state with its secondary states such as the irritation and disgust is investigated using the STFT parameters in Fig. 7.5. It shows the disgust state is more closure of the primary angry state as compared to the irritation state. The maximum magnitude of STFT of the angry state is −49.29 dB and of disgust state is −7.96 dB, whereas that of irritation state is 0.159 dB as observed from Fig. 7.5. Nevertheless, the difference in STFT magnitude of subsidiary disgust and irritation states is very meager indicating their similarity.

The cepstrum or log of the spectrum can approximate the ear more accurately. The high arousal cepstrum magnitudes have higher values due to higher frequency components as observed from Fig. 7.6.

A comparison of different statistical parameters of cepstrum has been made between different chosen emotions in Table 7.1 to fetch valuable information on the nature of affective states.

A comparison of the PSD among the primary angry state with its two subsidiary states such as the irritation and the disgust has been made in Fig. 7.7. There is a high correlation between the disgust and the irritation states as these are similar. From the plot, it can be inferred that a large part of the voice components is residing in the range between 0 to 1000 Hz.

The ACF has been used to find the relation between the similar type of emotions versus the neutral state. As shown in Fig. 7.8, the fear and nervous signals are correlated as the latter is a subsidiary state.

Male speakers found to have a lower pitch than either the female or children due to a longer vocal cord. Similarly, older people tend to have a lower pitch than younger adults. When speech is associated with affective states, the pitch is found to be lower with low arousal states such as bored or sad as compared to high-frequency states angry, fear, surprise, happiness, etc. It is highly correlated between similar emotions such as fear or nervousness as compared to neutral as observed from Fig. 7.9. It shows the nervous affective state is closely associated with the fear state. The pitch of fear state is defined in the range of 300–500 Hz, whereas that of

Table 7.1 Comparison of cepstrum statistics across the different affective states

Emotions	Maximum	Minimum	Mean	Variance	Range	Median
Happy	2.21	−0.265	−2.01E − 05	3.52E − 05	2.23	−5.30E − 06
Angry	2.32	−0.272	−1.77E − 05	6.40E − 05	2.59	−6.46E − 06
Bore	0.779	−0.314	4.11E − 07	1.99E − 05	1.09	−9.16E − 07
Sad	1.25	−0.204	−1.10E − 05	1.85E − 05	1.45	−2.10E − 06

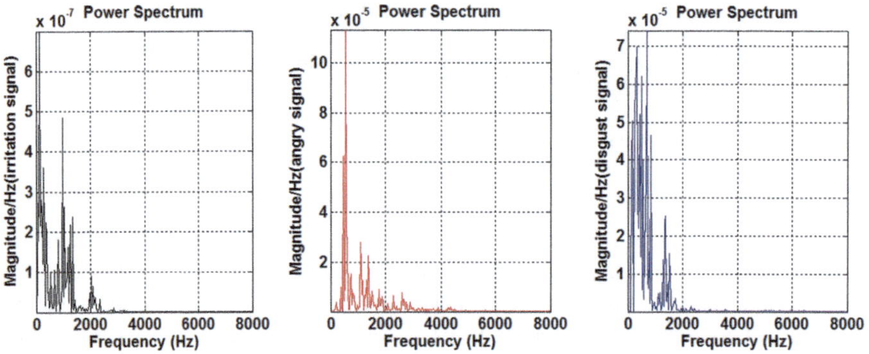

Fig. 7.7 A comparison among angry, irritation, and disgust voice samples using PSD

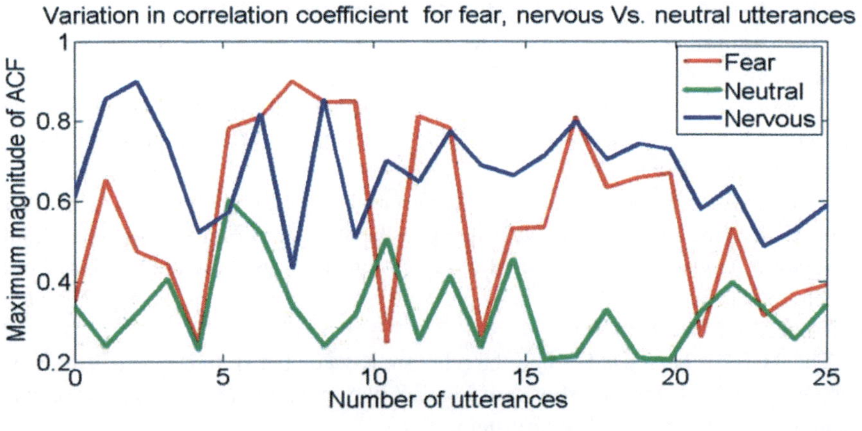

Fig. 7.8 Comparison of similar affective states using ACF

nervousness is in the range 200–300. Thus, it is possible to distinguish these states separately from that of the neutral state.

The ZCR indicates the number of times a voice sample crosses the zero axes. It is more when the fluctuations of a signal are rapid. Since high-arousal states contain higher intensity and frequencies, these have more ZCR than the low arousal states as shown in Fig. 7.10.

It is found that the speaker has a large speech rate during excited conditions than if he/she is in a cool mood. As observed from Fig. 7.11, the angry and happy states

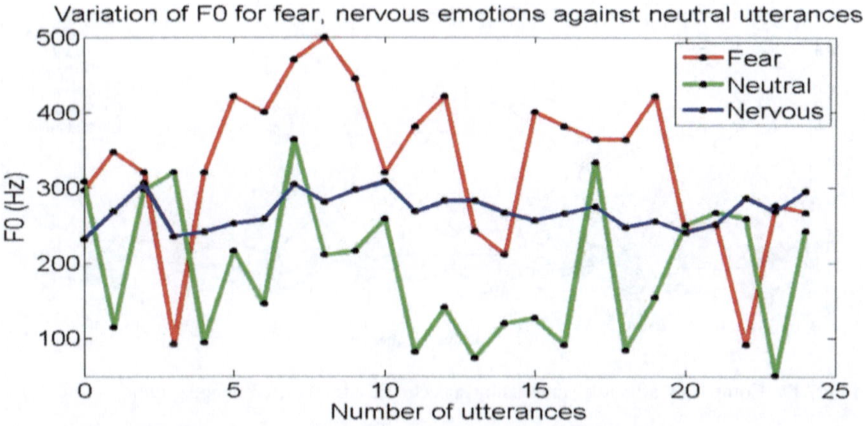

Fig. 7.9 The characterization of various human affective states using F0

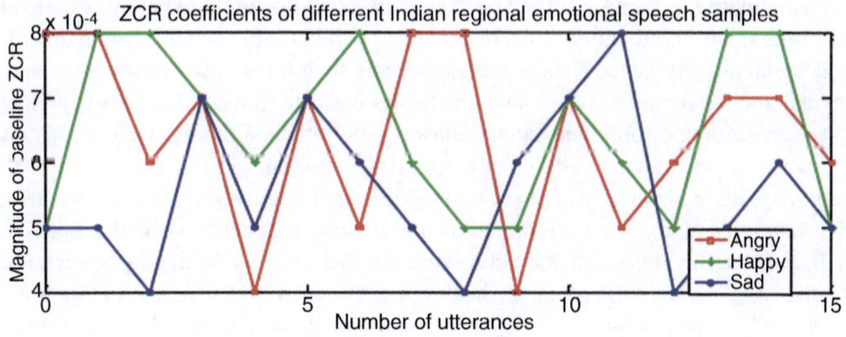

Fig. 7.10 The variation of ZCR among the different affective states

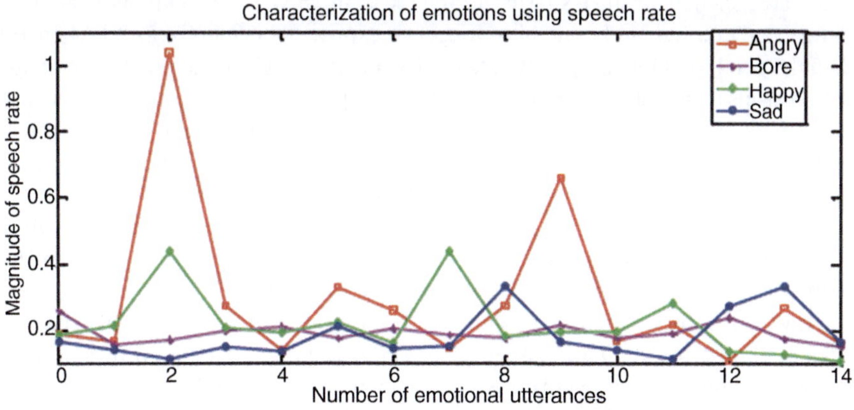

Fig. 7.11 Comparing affective states using speech rate across different age group

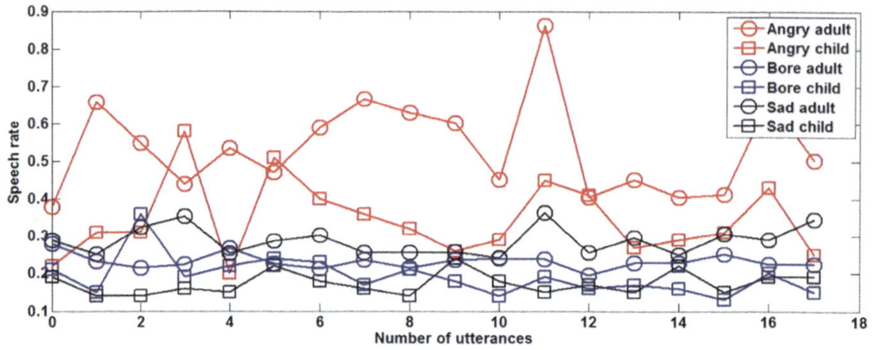

Fig. 7.12 Comparing affective states using speech rate across different age group

have the highest speech rate magnitudes as compared to either the sad or the bore affective states.

Similarly, the speech rate among speakers belonging to various age groups has been compared in Fig. 7.12. The speech rate of a child is found to be lower than that of adults as observed in Fig. 7.12. In general, a child is subjected to reading disorder and social anxiety hence speaks at a slower rate than the adults. A limited environmental and language exposure hampers the speech rate of a child as he/she ponders between suitable words/vocabulary during emotional outbursts. Further, a child arguably invents own words during speaking emotionally. On the contrary, the adults use learned words with associative skills and imagination under certain affective states. Thus, the reaction time of a child reduces that decrease his/her speaking rate. Further, the biological and neuromuscular factors are affecting his speech rate. As the child grows with age, he or she develops the desired oral-motor and linguistic skills such as the semantic, lexical, and phonological components. An improvement in the motor planning specificity of a grown-up child improves the articulation rate. This cognitive development as the child grows increases the speaking fluency.

The speech rate is found to be more with high-frequency states such as anger and fear state as a person speaks faster during these affective states with high energy. It indicates that a human being takes longer to express their moods with low energy than the corresponding aggressive states. The intensity of a signal remains high during an arousal state due to higher-frequency components. This is due to an increase in the strength of an utterance that rises involuntarily and automatically during an agitated or excited state. On the contrary, low energy or amplitude is manifested with a dull voice due to a low level of excitation.

The log energy values are higher for the high-arousal affective states such as happy and angry. On the contrary, the bore and sad states show the low energy due to low-frequency components as revealed from Fig. 7.13.

Among the children and the adults, the energy of the former is higher than the later as shown in Fig. 7.14. The level of enthusiasm and excitation of a child to abnormal situations or overexpressive behavior as compared to the judgmental adults may be the possible reason for this.

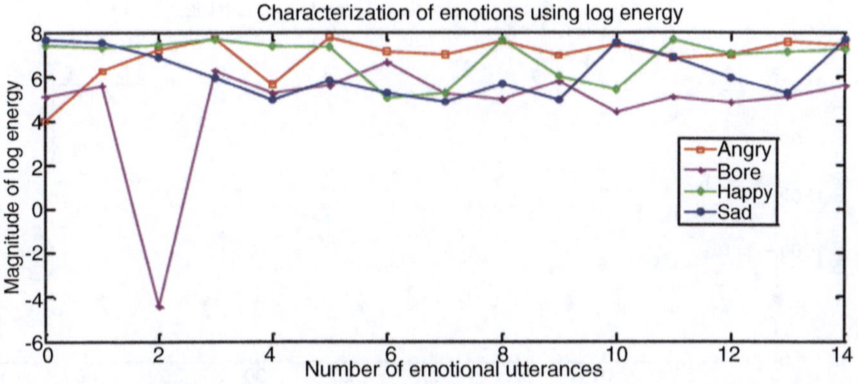

Fig. 7.13 The different affective states using log energy features

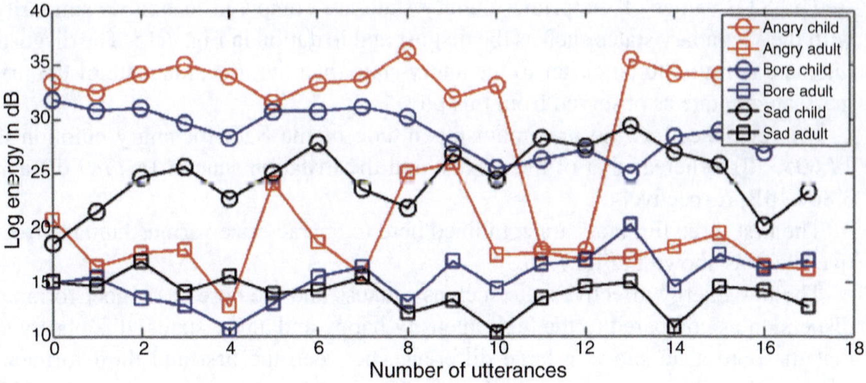

Fig. 7.14 The log-energy features characterizing the affective states across different age groups

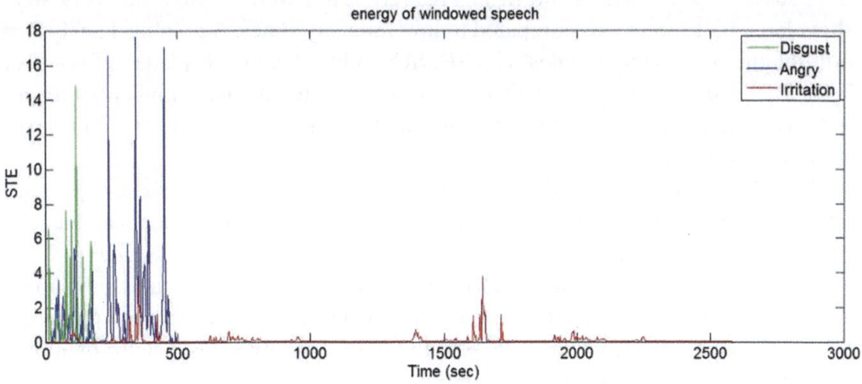

Fig. 7.15 The similar affective state such as angry, irritation, and disgust states using STE

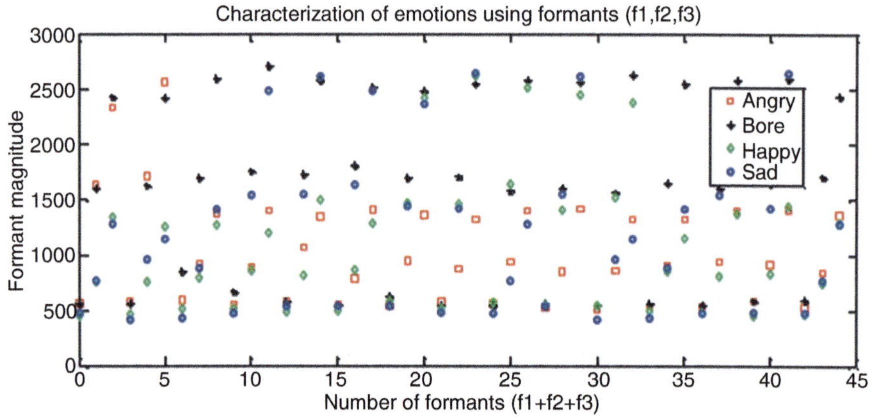

Fig. 7.16 Variation of formants among the different affective states

The STE values of the primary angry state are compared to find its similarity with the secondary states such as the disgust and irritation in Fig. 7.15. The disgusting state is found to be closer to the angry state than the irritation state of the primary angry state as observed from this plot.

It is evidenced as the maximum magnitude of the STE for angry emotion is 17.6053 dB, whereas that of the disgust and the irritation state is 14.7989 dB and 3.8072 dB, respectively.

The first three formants are combined here to characterize various human affective states as shown in Fig. 7.16.

The low-energy affective state such as sadness and the bore has higher formant dispersion as compared to the high intensity happy and angry states. It is observed that the bore state shows a large difference between the first and third formant, whereas the angry state shows the lowest. The formants have a range between 500 and 1700 kHz for the angry state, between 500 and 1500 kHz for the happy state, between 400 and 2600 kHz for the sad state, and between 500 and 2700 kHz for bore state, respectively, as shown in Fig. 7.16. However, the high-intensity states show large first formant as compared to low-intensity states. A graphical analysis of different human affective states with SR, SC, and SF features is made in Figs. 7.17, 7.18, 7.19 and 7.20. The angry, fear, and surprise states show the higher SR magnitude as compared to low-intensity emotional states due to higher spectral shape skewness as in Fig. 7.17.

The SC features are found to have a higher magnitude of surprise emotion followed by other high arousal states. It indicates the higher audio textures and level of maximum sound brightness with these high-intensity states as shown in Fig. 7.18.

A large local spectrum variation found with a surprise state indicates a maximum variation in the spectral energy distribution between two successive windows. A lower SF magnitude has been manifested with that of low arousal states than other high-arousal states as shown in Fig. 7.19.

The variation of the skewness magnitudes of 45 human voice samples has been plotted in Fig. 7.21. The objective is to compare two opposite human affective states

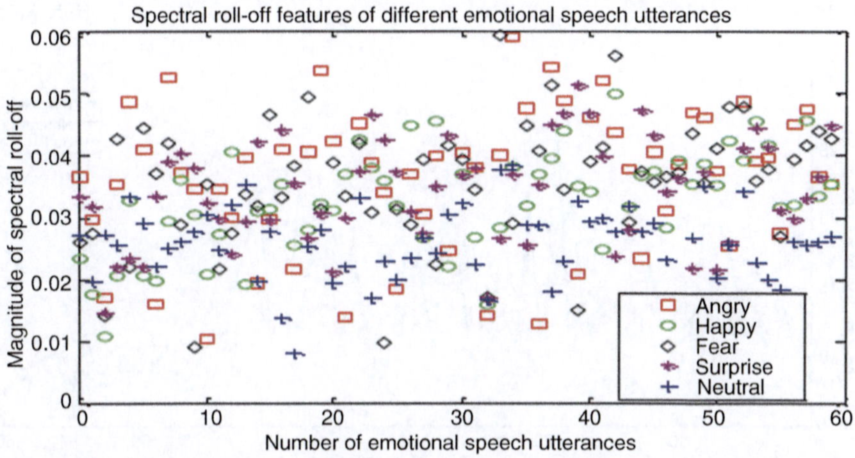

Fig. 7.17 The SR variation across different human affective states

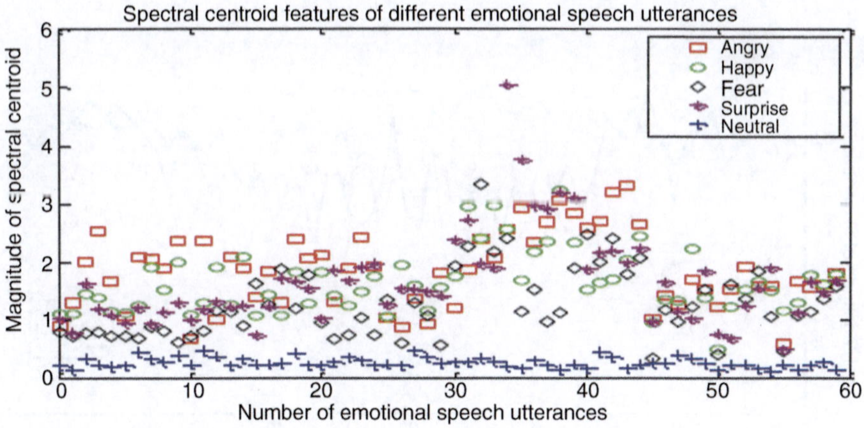

Fig. 7.18 The SC variation among the different human affective states

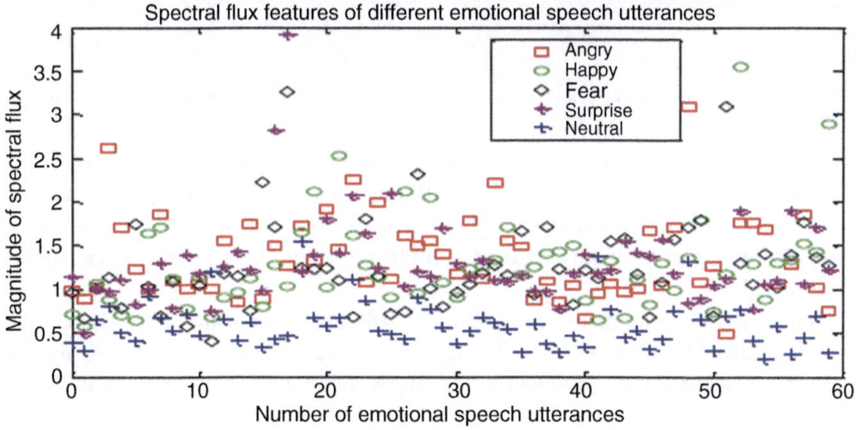

Fig. 7.19 The variation in SF among the different human affective states

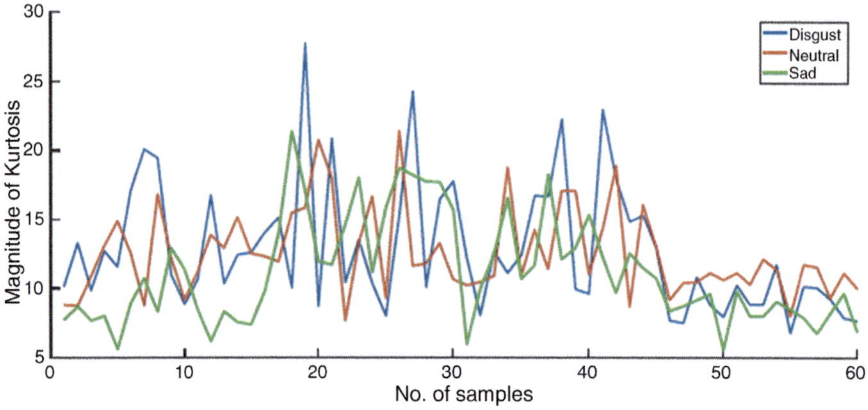

Fig. 7.20 The variation in kurtosis magnitude based on arousal states

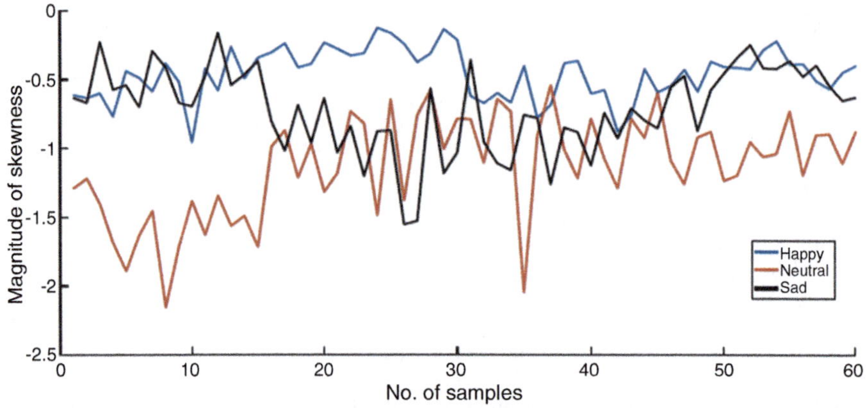

Fig. 7.21 The variation in skewness magnitude between two opposite states

such as happy and sad. The high-energy happy state shows the higher skewness values as compared to the low-frequency sad state as seen from Fig. 7.21.

Figure 7.22 provides the comparison of Shannon entropy values of 60 voice samples corresponding to two opposite affective states such as happy and sad. It is shown that the neutral state has similar entropy values to that of a low-arousal sad state. However, the happy being a high-frequency state has shown higher entropy values.

A comparison is made between different affective states using the average value of LPC coefficients in Fig. 7.23. The figure reveals a large variation in LPC magnitudes with surprise state followed by the angry and happy states. The ACF used in

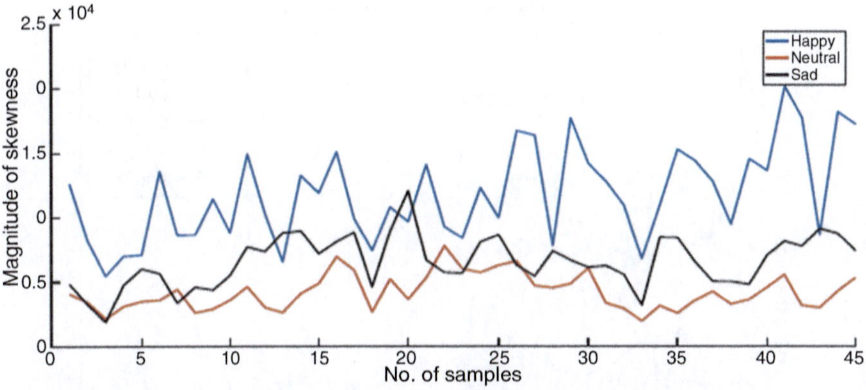

Fig. 7.22 The variation in Shannon entropy magnitude between two opposite states

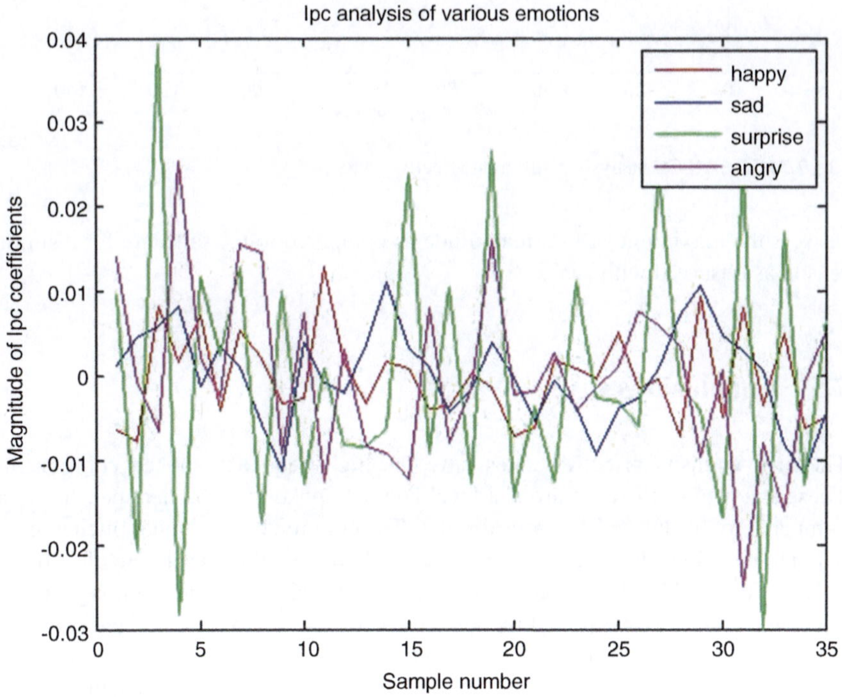

Fig. 7.23 LPC analysis of the human affective states

the extraction of LPCs signifies energy information, hence has the higher LPC magnitudes for the above-chosen states.

A similar comparison is made using the average value of MFCCs among different affective states in Fig. 7.24. In this case, the angry state has shown a large range of MFCCs variation followed by a happy state. Further, these states have shown to

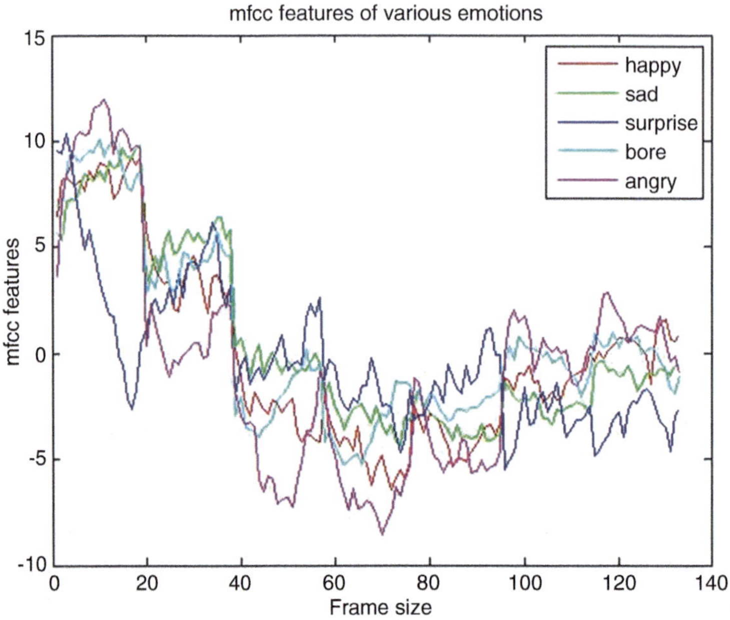

Fig. 7.24 The MFCC analysis of different affective states

possess the maximum MFCC magnitude as compared to the sad state for a similar reason as observed with LPCs.

7.5 Conclusions

The high-intensity affective states have the high-frequency spoken components. These states have the high-arousal level and are spoken at a higher speech rate as compared to the dull or low-arousal states. The children have a higher pitch as compared to the adults due to a longer vocal cord. However, the speech rate of a child is lower as the child is less aware of well-defined words and ponder among suitable vocabulary while expression of moods. The work also attempts to find the similarity among primary angry state with its subsidiary disgust and irritation states with different feature extraction techniques. A similar attempt is also made to demarcate the primary fear state with its secondary nervous state. The features have played a major role in characterizing different human affective states as the results reveal. It is possible to identify the chosen affective state across different age groups, arousal levels, and similar states.

References

Ai, H., Litman, D. J., Forbes-Riley, K., Rotaru, M., Tetreault, J., & Purandare, A. (2006). Using system and user performance features to improve emotion detection in spoken tutoring dialogs. In *Ninth International Conference on Spoken Language Processing*.

Bahreini, K., Nadolski, R., & Westera, W. (2016). Towards multimodal emotion recognition in e-learning environments. *Interactive Learning Environments, 24*(3), 590–605.

Frank, M. G., & Feeley, T. H. (2003). To catch a liar: Challenges for research in lie detection training. *Journal of Applied Communication Research, 31*(1), 58–75.

Hernández, A., Sanchez, V., Sánchez, G., Pérez, H., Olivares, J., Toscano, K., & Martinez, V. (2016, March). Security attack prediction based on user sentiment analysis of Twitter data. In *2016 IEEE International Conference on Industrial Technology (ICIT)* (pp. 610–617). New York: IEEE.

Kostoulas, T., Mporas, I., Kocsis, O., Ganchev, T., Katsaounos, N., Santamaria, J. J., & Fakotakis, N. (2012). Affective speech interface in serious games for supporting therapy of mental disorders. *Expert Systems with Applications, 39*(12), 11072–11079.

Martinez, R. S. (2017). *U.S. Patent No. 9,569,424*. Washington, DC: U.S. Patent and Trademark Office.

Merkx, P. A. B., Truong, K. P., & Neerincx, M. A. (2007, August). Inducing and measuring emotion through a multiplayer first-person shooter computer game. In *Proceedings of the Computer Games Workshop* (pp. 07–06).

Mohanty, M. N., & Palo, H. K. (2019). Segment based emotion recognition using combined reduced features. *International Journal of Speech Technology, 22*(4), 865–884.

Mohanty, M. N., & Palo, H. K. (2020). Child emotion recognition using probabilistic neural network with effective features. *Measurement, 152*, 107369.

Palo, H. K., & Behera, D. (2020). Analysis of speaker's age using clustering approaches with emotionally dependent speech features. In *Critical approaches to information retrieval research* (pp. 172–197). Hershey, PA: IGI Global.

Palo, H. K., Behera, D., & Rout, B. C. (2020). Comparison of classifiers for speech emotion recognition (SER) with discriminative spectral features. In *Advances in intelligent computing and communication* (pp. 78–85). Singapore: Springer.

Palo, H. K., Chandra, M., & Mohanty, M. N. (2017). Emotion recognition using MLP and GMM for Oriya language. *International Journal of Computational Vision and Robotics, 7*(4), 426–442.

Palo, H. K., Kumar, P., & Mohanty, M. N. (2017). Emotional speech recognition using optimized features. *International Journal of Research in Electronics and Computer Engineering, 5*(4), 4–9.

Palo, H. K., & Mohanty, M. N. (2018). Wavelet based feature combination for recognition of emotions. *Ain Shams Engineering Journal, 9*(4), 1799–1806.

Palo, H. K., & Mohanty, M. N. (2020). Analysis of speech emotions using dynamics of prosodic parameters. In *Cognitive informatics and soft computing* (pp. 333–340). Singapore: Springer.

Palo, H. K., Mohanty, M. N., & Chandra, M. (2016). Efficient feature combination techniques for emotional speech classification. *International Journal of Speech Technology, 19*(1), 135–150.

Palo, H. K., Mohanty, M. N., & Chandra, M. (2018). Speech emotion analysis of different age groups using clustering techniques. *International Journal of Information Retrieval Research (IJIRR), 8*(1), 69–85.

Palo, H. K., & Sarangi, L. (2020). Overview of machine learners in classifying of speech signals. In *Handbook of research on emerging trends and applications of machine learning* (pp. 461–489). Hershey, PA: IGI Global.

Paulmann, S., Bleichner, M., & Kotz, S. A. (2013). Valence, arousal, and task effects in emotional prosody processing. *Frontiers in Psychology, 4*, 345.

Petrushin, V. A. (2011). *U.S. Patent No. 7,940,914*. Washington, DC: U.S. Patent and Trademark Office.

Plutchik, R. (2001). The nature of emotions: Human emotions have deep evolutionary roots, a fact that may explain their complexity and provide tools for clinical practice. *American Scientist, 89*(4), 344–350.

Raj, J., & Kumar, S. (2015). Gender based affection recognition of speech signals using spectral & prosodic feature extraction. *International Journal of Engineering Research and General Science, 3*(2), 898–905.

Ram, R., Palo, H. K., & Mohanty, M. N. (2013). Emotion recognition with speech for call centres using LPC and spectral analysis. *International Journal of Advanced Computer Research, 3*(3), 189.

Song, K. T., Han, M. J., & Wang, S. C. (2014). Speech signal-based emotion recognition and its application to entertainment robots. *Journal of the Chinese Institute of Engineers, 37*(1), 14–25.

Tawari, A., & Trivedi, M. (2010, June). Speech based emotion classification framework for driver assistance system. In *2010 IEEE Intelligent Vehicles Symposium* (pp. 174–178). New York: IEEE.

Chapter 8
Embarrassment in the Context of Negative Emotions and Its Effects on Information Processing

Muskan Gumber, Anjali Sharma, and Salome Divya Joseph

Contents

8.1 Introduction

Self-conscious emotions, as the name implies, revolve around the self and the consciousness. These emotions comprise of the perception of us by self and the others, such as pride, shame, guilt and embarrassment (Else-Quest, Higgins, Allison, & Morton, 2012). There can be both positive and negative aspects to a self-conscious emotion which can further act as healthy motivators or inconvenient inhibitors. Embarrassment, usually characterized as a negative self-conscious emotion, is an unpleasant feeling that is a product of negative evaluation by self or others (Ayres, 2016).

M. Gumber · A. Sharma
CHRIST (Deemed to be University), Bengaluru, Karnataka, India

S. D. Joseph (✉)
SRM University, Amaravati, AP, India
e-mail: salome.d@srmap.edu.in

S. N. Mohanty (ed.), *Emotion and Information Processing*,
https://doi.org/10.1007/978-3-030-48849-9_8

This chapter aims to understand the emotion of embarrassment, its causes and consequences, particularly with respect to information processing. It further aims to explore gender differences in experience of negative emotions and how they vary in experience of embarrassment. Finally, this chapter proposes certain reasons for the identified variations in the experience of embarrassment and its effects.

8.2 Embarrassment and Embarrassability

Several attempts have been made to define embarrassment in terms of the kind of audience present in front of the individual or the factors that might influence the experience. According to Goffman (1956):

> embarrassment occurs whenever an individual is felt to have projected incompatible definitions of himself before those present. These projections do not occur at random or for psychological reasons but at certain places in a social establishment where incompatible principles of social organization prevail. (p. 264)

Similarly, Edelmann et al. (1987) identified social transgression and faux pas or accident as the factors leading to embarrassment. Another attempt at defining embarrassment has been made by Miller (2001) (as cited in Li, Yang, & Zhou, 2018) where he described it as:

> an individual's experience of an event that increases his or her likelihood of being subjected to an unwanted assessment, it can come from a real audience or only from an individual's imaginary audience, a panic, clumsy, distress-stricken, acute state of an individual that is embarrassing. (p. 688)

Embarrassability refers to a person's general susceptibility to embarrassment (Modigliani, 1968). This susceptibility varies from individual to individual, as some of them experience embarrassment to a great extent while for others, it is easier to overcome this feeling of discomfort.

8.3 Embarrassment and Related Emotions

Other than embarrassment, negative self-conscious emotions comprise of shame, guilt and humiliation. Even though these emotions are highly interrelated, the intensity and duration in which they are experienced, and the extent to which the self and the others are involved in each of them, differ.

Guilt, as defined in Collins English Dictionary, is an unhappy feeling arising out of the realization that one has done something wrong (Guilt, 2019). It is similar to embarrassment in a sense that both the emotions are products of wrong actions committed by people. However, guilt emerges when an individual realizes that his/her actions are wrong, and embarrassment occurs when others make one realize the same, either verbally or through non-verbal gestures.

Chambers Dictionary defines shame as a humiliating feeling occurring due to being unfavourable in front of the self or others (Shame, 2019). Where embarrassment represents more of one's social character, shame revolves around an individual's moral standards. Shame arises when individuals negatively evaluate themselves due to not living up to their own expectations. On the other hand, embarrassment results from being negative evaluated by others (Crozier, 2014). The major role played by self in the experiencing of shame is also supported by Pulham (2009) where shame is found to be more associated to self-caused events and humiliation and embarrassment to other-caused/accidental events.

According to Merriam-Webster Dictionary, humiliation refers to the act of reducing someone to a lower position in one's own eyes or others' eyes (Humiliation, 2019). The emotions of embarrassment and humiliation differ from each other mainly in terms of the intensity. Embarrassment, if given enough time, can be sublimed into a humorous anecdote. However, humiliation is traumatic and often cuts deeper (Burton, 2014a, 2014b). It was also found that overt demonstrations of rejection from others, such as laughter, demeaning looks and derogatory comments may be more common or necessary to elicit the emotion of humiliation (Pulham, 2009).

8.4 Embarrassment: Antecedents and Consequences

An interactional approach to understand a concept inculcates the biological, psychological and social factors revolving around it ("Psychological Disorders", 2015). It is a more convergent view that is based on the rationale that a single viewpoint is not sufficient to explore any condition in detail. This chapter uses this approach to identify the antecedents and consequences of experiencing embarrassment. In other words, the causes and outcomes of embarrassing experiences have been discussed in the light of biological, psychological, and social factors and its role in information processing.

8.4.1 Antecedents

The experience of embarrassment can arise from a number of factors. Sometimes, this experience can occur as the result of only one factor, whereas at other times, it is through the interaction of a few of these factors.

8.4.1.1 Biological

The experiencing of negative emotions is not solely based on the environment. Studies have found that the individuals carrying the short allele (s-allele) of the serotonin transporter gene have a higher tendency to react to negative emotions. Consequently, the carriers of the s-allele achieve higher scores for unpleasantness and goal-hindrance and lower scores for coping ability in response to fear and sadness, as compared to the non-carriers (Szily, Bowen, Unoka, Simon, & Kéri, 2008).

Brain differences also contribute to the differences in experiences of embarrassment. In 2011, Welsh identified 'the embarrassment centre' in the pregenual anterior cingulate cortex. This region is responsible for automatic bodily functions as well as thinking-related functions. The embarrassment centre is a tissue that resides deep inside the brain, to the front and the right. This region was found to be relatively smaller in people who experience low levels of embarrassment.

8.4.1.2 Psychological

Carl Rogers' contribution towards real and ideal self implies that inability to strive for an idealized version of the self leads to dissonance in an individual. A real self refers to what and who an individual actually is, whereas an ideal self is what an individual strives to become. This idealized aspect of the self emerges out of the expectations held by oneself and the others around them.

An incongruence between an individual's expectations from the self (ideal self) and how he/she actually acts in social settings (real self) ends up in the experience of embarrassment (Ayres, 2016).

8.4.1.3 Social

Certain theories have been proposed to emphasize on the positive and negative roles of society that leads to experiences of embarrassment.

Social Evaluation Theory, given by Goffman (1956), focussed on the erosion of esteem in the eyes of others as being important for an individual to experience eroded self-esteem and hence, embarrassment. This theory highlights the role of the social self of an individual, making them a passive recipient to the evaluation of others. In this light, individuals tend to view themselves the way others view them.

As cited in Sabini, Siepmann, Stein, and Meyerowitz (2000), Goffman (1967) also proposed the Dramaturgic Theory, wherein unless there is an actual or anticipated disruption of social performance, the mere thought of self or others' evaluation of us does not cause embarrassment. This theory varies from the Social Evaluation Theory as it places the individual in an active role where only if there is an actual or perceived disruptive act, he/she will experience embarrassment.

The revised version of the Social Evaluation Theory, provided by Modigliani (1971) (as cited in Robbins & Parlavecchio, 2006), argues that evaluation, both positive and negative, can result in the experience of embarrassment. Sabini et al. (2000) also contributed to this by proposing the Centre of Attention Theory which suggests that individuals can also experience embarrassment just by being the centre of attention. It is not necessary for them to face a loss of self-esteem or dramaturgic failure.

However, studies have also shown how individuals intentionally create embarrassing situations for others, in order to gain gratification. This type of embarrassment is called intentional embarrassment (Sharkey, 1992). The gratification sought can be extrinsic, intrinsic and/or social; extrinsic: to establish and/or maintain power; intrinsic: to experience pleasure, sense of pride and success, and self-satisfaction; social: to achieve an idealized role-identity in a social setting.

8.4.2 Outcomes

Just like the antecedents, experiencing embarrassment influences multiple components of human functioning, hence leading to the following consequences.

8.4.2.1 Biological

Müller-Pinzler et al. (2015) gave a model proposing the interaction between an individual's performance and how it is evaluated by the public. Negative evaluation of a low performance (failure) is a potential cause of embarrassment. On the basis of this model, Muller-Pinzler and his colleagues hypothesized that the interaction between failure and publicity can map a similar interaction between brain regions specific to mentalizing and the para-limbic system involved in human affect.

The mentalizing areas such as medial Prefrontal Cortex (mPFC) and precuneus are responsible for the processing of mental states, high-level cognitive functioning and self-related processing (Baetens, Ma, Steen, & Van Overwalle, 2013; Beer, Lombardo, & Bhanji, 2010; Margulies et al., 2009).There are significant cardiovascular responses to embarrassment Harris (2001); individual's blood pressure levels remain elevated even 5 min after embarrassing experiences.

Among other roles, the dorsal and ventral aspects of the anterior insula have an attentive and affective function, respectively (Wang et al., 2017). The amygdala, majorly has an emotional function and is involved in processing both positive and negative emotions (Anderson, 2007).

Thinking about how others will evaluate an individual on his/her low performance results in activation of the mentalizing areas (mPFC and precuneus). Low and high performance also activated the dAI, which showed an increased interaction with amygdala and vAI when an individual experienced the emotion of embarrassment (Müller-Pinzler et al., 2015).

Thus, it can be concluded that the experience of embarrassment has a neurobiological underpinning in the form of interaction between mentalizing areas (responsible for public feedback and social evaluation) and the para-limbic system (associated with processing of emotions).

8.4.2.2 Psychological

Embarrassment, being the negative evaluation of self, by self and the others, can have adverse psychological effects. It can make an individual question their self-worth and self-image, which can further result in a decrease in levels of self-confidence and self-esteem.

Depending on the specific cause and intensity of embarrassment in a situation, this experience can be a trigger for certain psychological conditions and maladaptive behaviours.

Being exposed to intense and frequent experiences of embarrassment may lead to health issues such as anxiety and depression. It can further result in the development of negative thought patterns and reduced self-worth potentially leading to disordered eating, social isolation, acts of self-harm, or thoughts of suicide ('Embarrassment', 2015). Schneier (1992) also reported that those who develop a fear of embarrassment may become socially withdrawn and depressed (as cited in Müller-Pinzler et al., 2015).

Embarrassment as an experience in a specific situation might end in a fraction of seconds, but its aftermath may persist for a considerably longer duration. As the individuals who have gone through experiencing embarrassment might relive it, this can be a possible explanation to why embarrassment can lead to the development of psychological conditions. However, given that the research evidence in this domain is almost negligible, more studies are required to determine how embarrassment and psychological conditions are related and whether treating embarrassment can also bring a positive change in an individual's mental health.

8.4.2.3 Social

An act that leads to embarrassment in one situation can be appropriate for another social setting. Hence, experiences of embarrassment result in role segregation where an individual specifies his or her role depending on the context. This acts as a possible way of reducing the chances of placing oneself in an embarrassing situation again. Role segregation gives way to audience segregation as an individual decides which kind of self is to be presented to which group of people (Goffman, 1956).

Studies have shown that people relate more with individuals who feel embarrassed as compared to people who display prideful facial expressions (Feinberg, Willer, & Keltner, 2012). This also contributes to how in cases of confrontations, showing embarrassment speeds up deescalation process because the person opposite to the embarrassed person feels bad about their behaviour and immediately withdraws from the argument (Ayres, 2016).

Feinberg et al. (2012) conducted a series of five experiments and their results showed that embarrassed individuals are more trustworthy, generous, prosocial and monogamous.

A possible explanation for these positive outcomes of experiencing embarrassment can be provided in the light of the Social Evaluation Theory (Goffman, 1956).

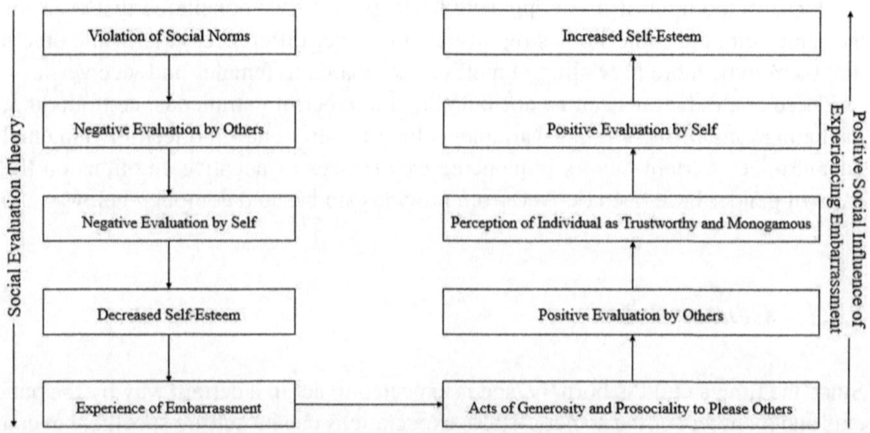

Fig. 8.1 A model to explain the positive social influence of experiencing embarrassment in the light of Social Evaluation Theory, proposed by Goffman (1956)

Given that individuals evaluate themselves negatively due to negative evaluation of self by others, there might be a tendency to change this negative evaluation into a positive one. This makes people act more generously, enhancing their abilities to become prosocial in nature. Generous and prosocial behaviour can be perceived positively by others. Individuals who are usually generous also gain a trustworthy standpoint from the society. When an individual is viewed in a positive light, he/she is more likely to sustain healthy and long-term relationships with others, hence, leading to monogamy. Positive evaluation of self by others can further change the earlier negative evaluation of self by self, eventually increasing one's level of self-esteem (Fig. 8.1).

8.5 Gender Differences in Experiencing Embarrassment

There is a plethora of evidence contributing to the emotional expression in individuals and whether or not, it varies according to their gender. Chaplin and Aldao (2013) found that positive emotion expressions and internalizing negative emotion expressions are more evident in girls, whereas externalizing negative emotions are expressed more by boys.

However, Ferguson and Eyre (2000) reported contradictory findings in gender differences based on previous evidence. For instance, they found that female preschoolers exhibit more overt aggression and self-distress, as components of guilt, whereas Chaplin and Aldao (2013) reported boys to express more externalizing emotion such as anger or aggression as compared to girls. Similarly, where some studies state that gender differences exist in the experiences of emotion, certain others claim that no differences can be observed till infancy.

Hence in the light of these opposing findings, it can be concluded that even for the same emotion, some studies report that no gender differences exist while others find them to be more prevailing in males as compared to females and vice versa.

These contradictions can be attributed to a number of extraneous, confounding, mediating, and/or moderating variables which either went unnoticed or remained uncontrolled. Certain factors influencing experiences of negative emotions on the basis of gender have been derived from previous studies and compiled below.

8.5.1 *Gender Roles*

Since the time a child is born, he/she is expected to act in a certain way by the parents and members of the society. These expectations can be culture specific, but can give rise to gender roles which are norms of conduct that has to be followed by individuals of a specific gender. Instead of one gender over the other, individuals with a feminine gender role reported more guilt-proneness and shame-proneness than did the masculine and undifferentiated gender role groups (Benetti-McQuoid & Bursik, 2005).

Gender in itself is a social construct, emerging out of the social roles or the need for a personal identity in an individual. As insensitivity is viewed as non-normative for females, they tend to develop feelings of guilt over inconsiderate behaviour whereas since males perceive aggression to be normative for them, they are less likely to show guilt over aggressive behaviours, as compared to females (Bybee, 1998). This indicates the importance of how individuals view themselves and how they are viewed by others, as an integral component in determining the manifestation of emotional experiences.

8.5.2 *Stereotypical Behaviours*

There are certain actions which seem to be accepted in one gender but not in the other. Ferguson and Eyre (2000) emphasize on the role of socialization agents to promote feelings of negative emotions of guilt and shame, more in females than males. Women are encouraged to be loving towards others, connected to them and interpersonally sensitive (Zahn-Waxler, Cole, & Barrett, 1991).

On the other hand, males are stereotyped to be achievement-oriented, active, aggressive, autonomous, competitive, dominant and stronger than females. This effectively provide few reasons for males to be ashamed of possessing unwanted identities. Plant, Hyde, Keltner, and Devine (2000) reported that women are stereotyped as experiencing more guilt, shame, and embarrassment, whereas men are stereotyped as experiencing more pride (as cited in Else-Quest et al., 2012).

Due to the complex interaction individuals have with the members of their society, it is likely to affect their thoughts, emotions and behaviours. Therefore, depending on what stereotypes are being followed in a given socio-cultural context, the emotional experiences may or may not give way to gender differences.

8.5.3 Age

The gender differences have been studied widely in different age groups of individuals. Certain meta-analytic studies provide evidence of how these differences increase with age. Chaplin and Aldao (2013) found that during infancy, gender differences were almost negligible but became more significant with age, such as in the period of adolescence. On the contrary, Else-Quest et al. (2012) reported that gender differences in experiencing guilt increase from adolescence through adulthood but found no significant gender differences in the experience of embarrassment.

One possible explanation for this result can be the gender roles and stereotypical behaviours mentioned above. As infants are unable to understand the complex expectations held by their parents and lack awareness about stereotypical behaviours around them, they are less likely to modify their behaviour along those lines. However, with age, the child begins to realize gender-specific roles and stereotypes which leads to the occurrence of gender differences. Another possible explanation is the change in emotion-related neural circuitry caused by hormonal levels at puberty which can lead to different emotional expression in individuals with different genders (De Bellis, 2001).

8.5.4 Self-Report

Studies have also been conducted to examine whether there is a difference in the way males and females report emotional experiences by themselves. Most of the findings conceptualize that females report these experiences more than males. It was found that more guilt is experienced in females, with the onset of adolescence. As cited in Bybee (1998), in a study conducted by Buss and Brock (1963), female college students reported more guilt, when asked to administer electric shocks, as compared to their male counterparts. Stilwell and Galvin (1985) also mentioned that girls reported feeling more discomfort and physiological disturbance in response to transgressions. In an empirical study conducted by Pulham (2009), the results indicated that females are more likely to experience embarrassment and humiliation than males.

A possible reason for these unidirectional findings may again be attributed to how females are expected to be more emotionally expressive than males. Parkins (2012) reported women to indeed be the more emotionally expressive of the genders. There also exist neural explanations to support that females experience more emotions than do males. In a study that required participants to use reappraisal to downregulate their emotional responses to negatively valenced pictures, the findings suggested that neural differences were evident between different genders, even when behavioural were not. Among the males, in terms of reappraisal, emotional responding and reward processing, a relatively lesser increase in prefrontal regions, a greater decrease in amygdala and lesser engagement of ventral striatal regions was found respectively (McRae, Ochsner, Mauss, Gabrielli, & Gross, 2008). These findings and explanations can be used to conclude that females, in general, have both an inherent tendency and a societal freedom to experience and report negative emotions, more than the males.

8.5.5 Situational Context

The intensity of negative emotions experienced and the gender differences in these experiences can be very situation-specific. For instance, an activity that is acceptable only for males to be performed in one situation might induce feelings of shame in females if they perform that action in that situation. Cradock O'Leary and Thurston (2012) found that females experience more shame with regard to physical appearance, relationship failure and academic failure, whereas males feel more ashamed when encountering failure in sports. With respect to guilt-proneness, women are more likely to mention guilt over lying and inconsiderate behaviour, whereas men report more guilt over not helping others, property damage, fighting, victimizing animals, aggressive and boisterous activities (Tangney, 1992; Williams & Bybee, 1994).

It was also found that boys exhibit more negative emotions than girls when they receive a disappointing gift as compared to an attractive gift and when they fail a game task (Chaplin & Aldao, 2013). Out of the ten dimensions that were studied in context of embarrassment, Else-Quest et al. (2012) found gender gaps to be largest for embarrassment about degrading the environment and smallest for embarrassment about sex. The underlying trend in the situations present above again points to the understanding that females experience more negative emotions when they either disappoint someone or perceive a threat to their relationship with others. On the other hand, males are more expressive of these emotions in situations which involve anger, aggression, violence and physical activity. Thus, it can be concluded that gender roles and stereotypical behaviours still have an influence, even when the situations in which emotions are being experienced may vary.

8.5.6 Interpersonal Context

Gender differences in the experiences of emotions is also influenced by interpersonal context as reported by Williams and Bybee (1994). According to their findings, while describing guilt-producing incidents, females made mention of family members and close friends, whereas men alluded casual acquaintances and people they are superficially involved with. Females were also more likely to mention parents as a source of guilt. In case of children, fewer gender differences are observed where they are with someone they know and trust well than when they are with an unfamiliar person.

Zeman and Garber (1996) found that children may behave in a socially acceptable manner in front of people they do not know. They are more likely to express negative emotions to those whom they know would be accepting such as parents and siblings rather than peers who might subject them to ridicule or rejection (Chaplin & Aldao, 2013). Thus, it can be said that the kind of relationships one shares with others and the amount of comfort present in those relationships also determines how the emotions are experienced and manifested.

To summarize, it can be stated that even though there is a lot of evidence to support that females experience negative emotions more than males, there still exist contradictions to these findings. The contrasting results can be attributed to the existence of the variables mentioned above. It can also be seen that not just one or two, but a complex interaction between all these variables is what leads to the emergence of gender differences in emotional experiences.

8.6 Experiencing Embarrassment Versus Experiencing Related Negative Emotions: Distinction Based on Gender Differences

Literature review indicates that experiences of embarrassment on the basis of gender, either do not exist at all or exist in relation to females experiencing them more than males. Where Else-Quest et al. (2012) found that gender differences in embarrassment were non-significant and negligible in magnitude, Pulham (2009) reported that women experience more embarrassment than males. Even when embarrassment is as negative an emotion as guilt, shame and humiliation, there is considerably less evidence to support the findings based on gender differences in its experience.

In the light of the two studies that have been reviewed with respect to gender differences in embarrassment, and factors mentioned above that might influence these differences in other negative emotions, there can be certain reasons to explain such findings.

Women are considered to be more emotionally expressive of the genders (Parkins, 2012). This finding is further supported by the neural differences that exist in the experience of negative emotions (McRae et al., 2008).

As discussed in Table 8.1, the involvement of self is less in experiencing embarrassment than in other negative emotions. This might explain why individuals report more of negative emotions such as shame and guilt but not embarrassment. Similarly, given that the nature of norms being violated in the case of embarrassment is more social than moral, and women are also expected to follow social norms more than males, it can be explained why women experience more embarrassment in causing environmental pollution than any other activity. Sex-related activities are not so socially explicit, which might explain why gender differences were smallest in this case.

In terms of intensity and duration, embarrassment is considered to be of a low intensity and less persistent than other negative emotions. Thus, it is possible that when an individual is asked to report the emotion being experienced or recall an embarrassing situation and mention the feelings being associated with it, he/she is more likely to report experiencing shame and guilt, than embarrassment.

8.6.1 Is Embarrassment as Simple an Emotion as It Is Considered to Be?

In light of the arguments put forth in this chapter, it can be deduced that there is a lot more to embarrassment as an emotion that goes beyond one's current understanding of it. This makes this emotion less 'negative' and more 'complex' in nature.

Every individual experiences embarrassment in a different manner. Instead of being static, common and always negative, these experiences can therefore be considered to lie on a continuum. On one end of this continuum are positive experiences and outcomes of embarrassment while on the other extreme, lie the completely negative experiences and outcomes that may further lead to certain psychological issues (Fig. 8.2).

Table 8.1 Similarities and differences in four self-conscious negative emotions

	Relevance in each of the following emotions			
Basis of distinction	Embarrassment	Shame	Guilt	Humiliation
Intensity	Low	High	Low	High
Duration	Less persistent	Highly persistent	Less persistent	Highly persistent
Involvement of self	Less	High	High	Less
Involvement of others	High	Less	Less	High
Nature of norm violation	More social; Less moral	More moral; Less social	More moral; Less social	More social; Less moral

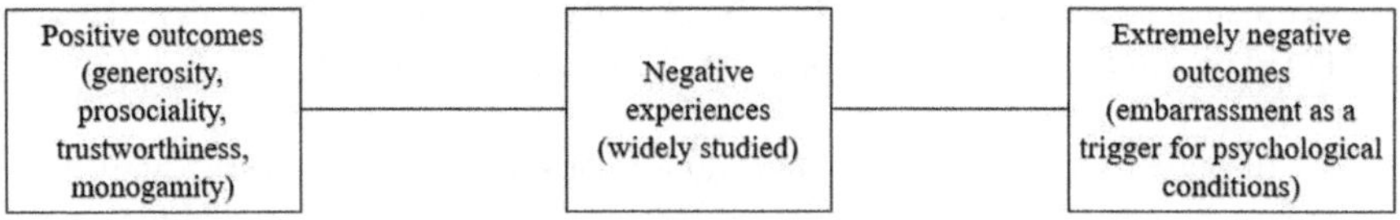

Fig. 8.2 The continuum of embarrassing experiences

8.7 Embarrassment and Information Processing

The interaction of bio-psycho-social factors in the experience of embarrassment bring us to the question: what role does the nature of embarrassment have on cognition or information processing?

Welsh (2011) identified 'the embarrassment centre' to be located at the pregenual anterior cingulate cortex. This particular region is also responsible for automatic bodily functions as well as thinking-related functions, particularly, decision-making. It is also established that immediate emotions can influence decision-making in different ways, by providing heuristic information to the decision maker (Schwarz & Clore, 1983). Individuals can make quick decisions that they would not otherwise make so easily: this is sometimes referred to as 'presence of mind'. Embarrassment could influence appraisals of subsequent information and events (Tiedens & Linton, 2001). It has been found that in the process of decision-making, positive affectivity increases reliance on heuristic processing, i.e. using heuristic cues; whereas negative affectivity is more linked with the tendency to involve in systematic processing (Mohanty & Suar, 2013a; 2014; Mohanty, Pratihar, & Suar, 2015). Embarrassment, as an emotion is more prone to causing negative affectivity, and therefore, can result in the application of systematic processing during decision-making. This can serve as a form of priming. Thereby cognitively preparing the individual for future responses.

Embarrassment can motivate certain types of stimulating or inhibiting behaviour (Raghunathan & Pham, 1999). A state of negative mood, such as the one induced by an experience of embarrassment, can increase the response time for decision-making and slow down the process of thinking (Mohanty & Suar, 2013b; Mohanty et al., 2015). On, the other hand, embarrassment associated with sad feelings can also make individuals take decisions that make them feel better, like gifting themselves something or makings choices that will improve levels of happiness. Similarly, when embarrassment is linked to anxiety, it can lead to behaviours that avoid risk or at least reduce uncertainty. However, quite the opposite can be observed when embarrassment is linked to shame. It can produce avoidant behaviour like failing to seek care for many types of medical conditions (Consedine, Krivoshekova, & Harris, 2007; Harris, 2006; Shinn et al., 2004) and can also lead people to engage in risky sexual practices such as failing to obtain and use condoms (Leary & Dobbins, 1983; Moore, Dahl, Gorn, & Weinberg, 2006). It has also been found that individuals can make people neglect helping others who may be in serious need, in order to avoid an embarrassing situation (Sabini, Siepmann, & Stein, 2001) (Fig. 8.3).

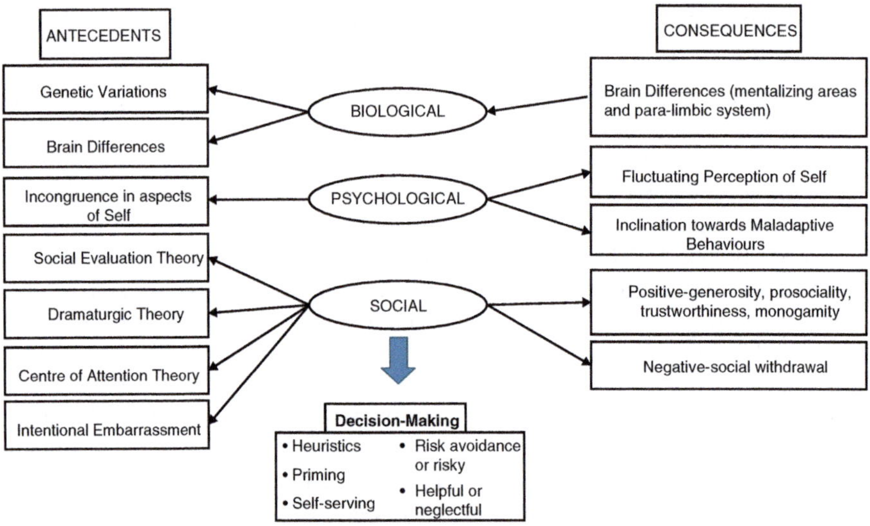

Fig. 8.3 Summary of antecedents and consequences of experiencing embarrassment, using the interactional (bio-psycho-social) approach, and its effect on information processing

8.8 Placing Individuals on the Embarrassment Continuum

Tracing the aforementioned antecedents, consequent and gender differences of experiencing negative emotions, three broad determinants that may place a particular experience on the proposed continuum have been identified.

8.8.1 Personal Factors of an Individual

The individual's emotional response to an embarrassment-provoking situation depends on their genetic makeup and general perception of self. These factors include whether a person is a carrier of the s-allele of the serotonin transporter gene, the size of the embarrassment centre in an individual's brain, level of self-confidence and self-esteem, and so on.

8.8.2 Nature of the Situation

The response to experiences of embarrassment is also influenced by the event that causes it (such as violence, insult) and the event in which it is caused (such as in a social gathering). The individual's familiarity with the people witnessing the embarrassing event also plays a role.

8.8.3 *Social Support*

The way people perceive an individual's behaviours, the expectations they hold from them and the evaluation they make of the individual's embarrassing experiences also contribute in placing an experience on the continuum.

8.9 Limitations

Firstly, the conceptualizations are solely on the basis of previous studies and their findings. Secondly, more literature was based on the western as compared to the eastern context. Since the socio-cultural stereotypes might be an integral component, research based on emotional experiences, specific to eastern context, might bridge this gap.

8.10 Implications and Future Scope

Considering the emphasis placed on embarrassment as an emotion and comparisons that were made between embarrassment and its related negative emotions, it can be concluded that embarrassment has a lot of undiscovered facets to it, and hence, it is not as simple an emotion, as it might be considered. The complexity of this emotion can be attributed to a variety of factors. The experience of embarrassment is largely viewed with a negative conception but as the social outcomes discussed above reveal, individuals who feel more embarrassed are also likely to be generous, prosocial, trustworthy and monogamous. Thus, experiencing embarrassment can also have a positive connotation attached to it which needs to be further explored. Likewise, strategies to control the negative effects of embarrassment in terms of poor decision-making, need to be evolved.

In terms of gender differences, it was observed that females experienced negative emotions, including embarrassment, more than males. However, in the light of contradictory findings, it is necessary that more research is carried out in this domain to have a clear understanding of gender differences in emotional experiences. In order to examine the influence of factors like socio-cultural stereotypes and situational context, cross-cultural studies can be conducted. Similarly, longitudinal studies can be carried out to gain further clarity of age as a moderating variable, in the experience of emotions based on gender differences.

Furthermore, even though the existing evidence to support this claim is considerably negligible, there might be a possibility that the relationship between psychological conditions and embarrassment is bidirectional. This means that as much as psychological conditions lead to experiences of embarrassment, they can also get triggered by these experiences. Further research in this area can provide substantial findings with regard to this proposition.

Finally, the proposed conceptual framework of the continuum of embarrassing experiences requires empirical support to be theorized as a model. Both qualitative and quantitative studies can be carried out to explore the range of embarrassing experiences individuals come across and the factors attributed to them.

8.11 Conclusion

This chapter explored embarrassment, its causes and consequences, particularly with respect to information processing. It further reviewed gender differences in the experience of negative emotions and how they vary in experience of embarrassment. Finally, this chapter proposed certain reasons for the identified variations in the experience of embarrassment and its effects.

References

Anderson, A. K. (2007). Feeling emotional: The amygdala links emotional perception and experience. *Social Cognitive and Affective Neuroscience, 2*(2), 71–72.

Ayres, A. (2016, January 9). *Why we get embarrassed and how to overcome it.* Retrieved from https://lifehacker-com.cdn.ampproject.org/v/s/lifehacker.com/why-we-get-embarrassed-and-how-to-overcome-it

Baetens, K., Ma, N., Steen, J., & Van Overwalle, F. (2013). Involvement of the mentalizing network in social and non-social high construal. *Social Cognitive and Affective Neuroscience, 9*(6), 817–824. https://doi.org/10.1093/scan/nst048

Beer, J. S., Lombardo, M. V., & Bhanji, J. P. (2010). Roles of medial prefrontal cortex and orbitofrontal cortex in self-evaluation. *Journal of Cognitive Neuroscience, 22*(9), 2108–2119.

Benetti-McQuoid, J., & Bursik, K. (2005). Individual differences in experiences of and responses to guilt and shame: Examining the lenses of gender and gender role. *Sex Roles, 53*(1). https://doi.org/10.1007/s11199-005-4287-4

Burton, N. (2014a, August 26). The psychology of embarrassment, shame, and guilt. *Psychology Today.* Retrieved from https://www.psychologytoday.com/us/blog/hide-and-seek/201408/the-psychology-embarrassment-shame-and-guilt

Burton, N. (2014b, August 27). The psychology of humiliation. *Psychology Today.* Retrieved from https://www.psychologytoday.com/us/blog/hide-and-seek/201408/the-psychology-humiliation

Bybee, J. (1998). The emergence of gender differences in guilt during adolescence. In *Guilt and children* (pp. 113–125). Cambridge, MA: Academic Press. https://doi.org/10.1016/b978-012148610-5/50006-0

Chaplin, T. M., & Aldao, A. (2013). Gender differences in emotion expression in children: A meta-analytic review. *Psychological Bulletin, 139*(4), 735–765. https://doi.org/10.1037/a0030737

Consedine, N. S., Krivoshekova, Y. S., & Harris, C. R. (2007). Bodily embarrassment and judgment concern as separable factors in the measurement of medical embarrassment: Psychometric development and links to treatment-seeking outcomes. *British Journal of Health Psychology, 12*, 439–462.

Cradock O'Leary, J., & Thurston, N. S. (2012). *Gender differences in experiencing shame.* Faculty Publications – Grad School of Clinical Psychology, Paper 12. Retrieved from http://digitalcommons.georgefox.edu/gscp_fac/12

Crozier, W. R. (2014). Differentiating shame from embarrassment. *Emotion Review, 6*(3), 269–276. https://doi.org/10.1177/1754073914523800

De Bellis, M. D. (2001). Developmental traumatology: The psychobiological development of maltreated children and its implications for research, treatment, and policy. *Development and Psychopathology, 13*(3), 539–564. https://doi.org/10.1017/s0954579401003078

Edelmann, R. J., Asendorpf, J., Contarello, A., Georgas, J., Villanueva, C., & Zammuner, V. (1987). Self-reported verbal and non-verbal strategies for coping with embarrassment in five European cultures. *Social Science Information, 26*(4), 869–883.

Else-Quest, N. M., Higgins, A., Allison, C., & Morton, L. C. (2012). Gender differences in self-conscious emotional experience: A meta-analysis. *Psychological Bulletin*. https://doi.org/10.1037/a0027930

Embarrassment. (2015, December 14). Retrieved from https://www.goodtherapy.org/blog/psychpedia/embarrassment

Feinberg, M., Willer, R., & Keltner, D. (2012). Flustered and faithful: Embarrassment as a signal of prosociality. *Journal of Personality and Social Psychology, 102*(1), 81–97. https://doi.org/10.1037/a0025403

Ferguson, T., & Eyre, H. (2000). Engendering gender differences in shame and guilt: Stereotypes, socialization, and situational pressures. *Gender and emotion: Social psychological perspectives*. Retrieved from https://www.researchgate.net/publication/312979157

Goffman, E. (1956). Embarrassment and social organization. *American Journal of Sociology, 62*(3), 264–271. Retrieved from https://www.jstor.org/stable/2772920

Guilt. (2019). *Collins English dictionary*. Retrieved from https://www.collinsdictionary.com/dictionary/english/guilt

Harris, C. R. (2001). Cardiovascular responses of embarrassment and effects of emotional suppression in a social setting. *Journal of Personality and Social Psychology, 81*(5), 886–897.

Harris, C. R. (2006). Embarrassment a form of social pain. *American Scientist, 94*(6), 524–533.

Humiliation. (2019). *Merriam-Webster dictionary*. Retrieved from https://www.merriam-webster.com/dictionary/humiliate

Leary, M. R., & Dobbins, S. E. (1983). Social anxiety, sexual behavior, and contraceptiveuse. *Journal of Personality and Social Psychology, 45*, 1347–1354.

Li, Y., Yang, D., & Zhou, H. (2018). A literature review of consumption embarrassment and prospects. *American Journal of Industrial and Business Management, 8*, 686–699. https://doi.org/10.4236/ajibm.2018.83046

Margulies, D. S., Vincent, J. L., Kelly, C., Lohmann, G., Uddin, L. Q., Biswal, B. B., … Petrides, M. (2009). Precuneus shares intrinsic functional architecture in humans and monkeys. *Proceedings of the National Academy of Sciences, 106*(47), 20069–20074. https://doi.org/10.1073/pnas.0905314106

McRae, K., Ochsner, K. N., Mauss, I. B., Gabrielli, J. J. D., & Gross, J. J. (2008). Gender differences in emotion regulation: An fMRI study of cognitive reappraisal. *Group Processes & Intergroup Relations, 11*(2), 143–162. https://doi.org/10.1177/1368430207088035

Modigliani, A. (1968). Embarrassment and embarrassability. *Sociometry, 31*(3), 313–326. Retrieved from http://www.jstor.org/stable/2786616

Mohanty, S. N., Pratihar, D. K., & Suar, D. (2015). Study on influence of mood states on information processing during decision making using fuzzy reasoning tool and neuro-fuzzy system developed based on Mamdani approach. *International Journal of Fuzzy Computing Modeling, 1*(3), 252–268. Retrieved from https://www.inderscience.com/info/inarticle.php?artid=699

Mohanty, S. N. (2014). Decision making under uncertainty and information processing in positive and negative mood states. *Psychological Reports, 115*(1), 44–64. https://doi.org/10.2466%2F20.04.PR0.115c16z

Mohanty, S. N., & Suar, D. (2013a). Decision-making in positive and negative prospects: Influence of certainty and affectivity. *International Journal of Advances in Psychology, 2*(1), 19–28. Retrieved from https://drsachinandan.com/publication/27.DM_IJAP.pdf

Mohanty, S. N., & Suar, D. (2013b). Influence of mood states, group discussion, and interpersonal comparison on change in decision-making and information processing. *International Journal of Decision Science, Risk and Management, 5*(2) 101–123. Retrieved from https://ideas.repec.org/a/ids/ijdsrm/v5y2013i2p101-123.html

Mohanty, S. N., & Suar, D. (2014). Influence of Mood States on Decision Making under Uncertainty and Information processing. *Psychological Reports, 115*(1), 44–64.

Moore, S. G., Dahl, D. W., Gorn, G. J., & Weinberg, C. B. (2006). Coping with condom embarrassment. *Psychology, Health & Medicine, 11*, 70–79.

Müller-Pinzler, L., Gazzola, V., Keysers, C., Sommer, J., Jansen, A., Frässle, S., … Krach, S. (2015). Neural pathways of embarrassment and their modulation by social anxiety. *NeuroImage, 119*, 252–261. https://doi.org/10.1016/j.neuroimage.2015.06.036

Parkins, R. (2012). Gender and emotional expressiveness: An analysis of prosodic features in emotional expression. *Griffith Working Papers in Pragmatics and Intercultural Communication, 5*(1), 46–54.

Psychological Disorders. (2015). *Psychology, textbook XII* (pp. 69–88). New Delhi, India: National Council of Education Research and Training.

Pulham, D. J. (2009). *Humiliation and its relationship to embarrassment and shame*. Electronic Theses and Dissertations. 905. Retrieved from https://digitalcommons.du.edu/etd/905

Raghunathan, R., & Pham, M. T. (1999). All negative moods are not equal: Motivational influences of anxiety and sadness on decision making. *Organizational Behavior and Human Decision Processes, 79*, 56–77.

Robbins, B. D., & Parlavecchio, H. (2006). The unwanted exposure of the self: A phenomenological study of embarrassment. *The Humanistic Psychologist, 34*(4), 321–345. Retrieved from https://psycnet.apa.org/buy/2006-23327-003

Sabini, J., Siepmann, M., & Stein, J. (2001). The really fundamental attribution error insocial psychological research. *Psychological Inquiry, 12*, 1–15.

Sabini, J., Siepmann, M., Stein, J., & Meyerowitz, M. (2000). Who is embarrassed by what? *Cognition and Emotion, 14*(2), 213–240. https://doi.org/10.1080/026999300378941

Schwarz, N., & Clore, G. L. (1983). Mood, misattribution, and judgments of well-being: Informative and directive functions of affective states. *Journal of Personality and Social Psychology, 45*(3), 513–523.

Shame. (2019). *Chambers 21st century dictionary*. Retrieved from https://chambers.co.uk/search/?query=shame&title=21st

Sharkey, W. F. (1992). Use and responses to intentional embarrassment. *Communication Studies, 43*(4), 257–275. https://doi.org/10.1080/10510979209368377

Shinn, E., Basen-Engquist, K., Le, T., Hansis-Diarte, A., Bostic, D., Martinez-Cross, J., … Follen, M. (2004). Distress after an abnormal pap smear result: Scale development and psychometric validation. *Preventive Medicine, 39*, 404–412.

Stilwell, B. M., & Galvin, M. (1985). Conceptualization of conscience in 11–12-year-olds. *Journal of the American Academy of Child Psychiatry, 24*(5), 630–636. https://doi.org/10.1016/S0002-7138(09)60066-2

Szily, E., Bowen, J., Unoka, Z., Simon, L., & Kéri, S. (2008). Emotion appraisal is modulated by the genetic polymorphism of the serotonin transporter. *Journal of Neural Transmission, 115*(6), 819–822. https://doi.org/10.1007/s00702-008-0029-4

Tangney, J. P. (1992). Situational detenninants of shame and guilt in young adulthood. *Personality and Social Psychology Bulletin, 18*(2), 199–206. https://doi.org/10.1177/0146167292182011

Tiedens, L. Z., & Linton, S. (2001). Judgment under emotional certainty and uncertainty: The effects of specific emotions on information processing. *Journal of Personality and Social Psychology, 81*(6), 973–988.

Wang, Y., Zhu, L., Zou, Q., Cui, Q., Liao, W., Duan, X., & Chen, H. (2017). Frequency dependent hub role of the dorsal and ventral right anterior insula. *NeuroImage*. https://doi.org/10.1016/j.neuroimage.2017.10.004

Welsh, J. (2011, April 15). *Embarrassed? Blame your brain*. Retrieved from https://amp-livescience-com.cdn.ampproject.org/v/s/amp.livescience.com/13727-embarrassment-brain-dementia.html?usqp=mq331AQCCAE%3D&amp_js_v=0.1#referrer=https%3A%2F%2Fwww.google.com&amp_tf=From%20%251%24s&ampshare=https%3A%2F%2Famp.

livescience.com%2F13727-embarrassment-brain-dementia.html%23referrer%3Dhttps%253A%252F%252Fwww.google.com%26amp_tf%3DFrom%2520%25251%2524s

Williams, C., & Bybee, J. (1994). What do children feel guilty about? Developmental and gender differences. *Developmental Psychology, 30*(5), 617–623. https://doi.org/10.1037/0012-1649.30.5.617

Zahn-Waxler, C., Cole, P. M., & Barrett, K. C. (1991). Guilt and empathy: Sex differences and implications for the development of depression. In J. Garber & K. A. Dodge (Eds.), *Cambridge studies in social and emotional development. The development of emotion regulation and dysregulation* (pp. 243–272). New York, NY: Cambridge University Press. https://doi.org/10.1017/CBO9780511663963.012

Zeman, J., & Garber, J. (1996). Display rules for anger, sadness, and pain: It depends on who is watching. *Child Development, 67*(3), 957–973. https://doi.org/10.2307/1131873

Chapter 9
Student's Emotion: The Power of Emotion Education in School

Auditi Pramanik and Swati Dhir

Contents

9.1 Introduction

This chapter puts light on the role of education on the emotional development of children and how these two aspects are intertwined. Emotion in educational contexts has obtained increasing research attention over the past 20 years because classroom environments are full of emotions, and emotions are considered an integral part of personal well-being and health. We can find researches on teacher's emotional experiences, children's emotional development through education, developing students' emotional competence, even studies that are focusing on the relationship between emotion and learning and achievement. Nowadays, some popularly used concepts such as 'emotional literacy' and social-emotional learning' highlight the

A. Pramanik (✉)
Jawaharlal Nehru University, New Delhi, India

S. Dhir
IMI New Delhi, New Delhi, India
e-mail: swati.dhir@imi.edu

S. N. Mohanty (ed.), *Emotion and Information Processing*,
https://doi.org/10.1007/978-3-030-48849-9_9

existence of the wide agreement on the importance of emotional education. It is a well-established fact that being in school or in the classroom, either as a student or as a teacher, can evoke a multitude of emotions. For instance, teachers and students both can feel proud after success, anxious before challenging tasks, angry when things are not turning out as they expected and so on. A classroom represents a highly interactive and emotionally charged setting, yet emotions in academic settings were largely neglected by educators and policymakers. Although studies have shown that students' emotions are related to their learning strategies, their motivation and ultimately their achievement (Linnenbrink, 2007; Mega, Ronconi, & De Beni, 2014; Ranellucci, Hall, & Goetz, 2015). Given this assumption, children need a combination of intellectual skills, motivational qualities, and socio-emotional skills to succeed in life. But unfortunately, many students have experienced that the current education system and its curriculum do not fully support their social and emotional development. Along with this children's mental health issues such as to conduct,[1] hyperkinetic,[2] emotional disorder,[3] and stress is of international concern. In the General Assembly of the United States, a resolution was adopted with its post-2015 Agenda where 17 Sustainable Development Goals were set out (United Nations General Assembly, 2015). Among those goals, two are directly about children's health and education. Particularly Goal three aims to "ensure healthy lives and promote well-being for all at all ages" (United Nations General, 2015, p. 14). These goals on health and education are very much intertwined because of the prominent role of education on children's overall development. This chapter is dedicated to understanding the critical relationship of education and emotion under the following headings:

- Role of education in children's emotional development
- Emotional attachment, cognition, and behavior
- Emotional coaching strategy for promoting sustainable emotional and behavioral well-being
- Teacher–students attachment and emotional development
- Emotional competence, emotional intelligence, and emotional literacy

[1] The ICD-10 characterizes conduct disorder as "a repetitive and persistent pattern of dissocial, aggressive, or defiant conduct," explaining that such behavior is "more severe than ordinary childish or adolescent rebelliousness" (World Health Organization, 1992, p. 209). Classified in a *DSM-5* chapter with disruptive and impulsive disorders, conduct disorder manifests in behavioral symptoms, such as aggressive or deceitful acts, that "violate the rights of others or that violate major societal norms" (p. 461).

[2] According to ICD-10, the hyperkinetic disorder is evident by "a combination of overactive, poorly modulated behavior with marked intention and lack of persistent task involvement; and pervasiveness over situations and persistence over time of these behavioral characteristics" (p. 206). The ICD-10 stressed that "[hyperkinetic] disorders always arise early in development (usually in the 5 years of life)" (p. 206).

[3] Emotional disorders describe feelings of distress, including anxiety and depression. The ICD-10 describes emotional disorders diagnosed during childhood as "exaggerations of normal developmental trends rather than phenomena that are qualitatively abnormal in themselves" (World Health Organization, 1992).

- Emotion and attitude development
- Social-emotional development and cooperative learning
- Importance of teacher's training program
- Emotional labor and education

At the end of this chapter, we would hope that the reader would be clearer about the critical and dual relationship of education and emotion and how both of them have impacted each other both positively and negatively.

9.2 Role of Education in Children's Emotional Development

Development involves changes within individuals and the dynamic exchanges between the individual and the environment over time. Being a social entity human beings are immersed in a particular culture and caregiving context is expanding, transforming, and recognizing with others is the unique way that carves out individual pathways for development. Notably, there are several reasons for researchers and educators to be concerned with emotional development because primarily, emotions help to define one's individuality. Current information-processing models (Greenwald et al., 2000; Medrano et al., 2015) indicate that self-concept consists of an enormous repertoire of relevant information about oneself that includes episodic as well as semantic knowledge. The working self-concept and the complete repertoire are organized in a series of categories, commonly known as self-aspects. These self-aspect categories are ideographic and, as previously mentioned, reflect individual roles in life in a specific situation (with friends, teachers, family), emotional states, personal traits, and interpersonal relationships, among others. Bower's Associative Network Model (Bower & Forgas, 2001) proposes that mood can cause a change in the activation of the categories of self-concept in any given moment. Henceforth, a change in mood or one's emotion is accompanied by a change in self-concept (Mohanty & Suar, 2014). Therefore, development is concerned with developing capacities for emotional expression, sociability, self-understanding, self-awareness, self-management, and other facets of socio-emotional growth. Recent researchers have proven that developing these core life abilities through social and emotional learning is critical to a child's development and school is the only place where any deficiencies in these abilities can be addressed before they become active members of society (Davis, 2003; Shaughnessy, 2012). Combining emotional skills with academic development creates high-quality learning experiences and environments that empower students to be more effective contributors in their classrooms today and in their workplace and communities tomorrow. According to the Allen Report (2011), "By building a child's social and emotional capabilities we enable children to be happily engaged with others and with society, and to learn, to develop fully, to attain and to achieve. In essence, it delivers school ready, life ready, and child ready members of society." Emotional attachment is crucial to children's psychological welfare and forms the basis of personality development and socialization (Bowlby, 1988a, 1988b). Teachers, parents, and significant others in a child's life can provide important attachments for children (Bergin & Bergin, 2009; Riley, 2010). Recent advances in neuroscience have highlighted the connections between emotion, social functioning, and decision-making that have the potential to revolutionize our understanding of the role of emotion in education and vis-a-vis. In particular, the neurobiological evidence suggests that the aspects of cognition that we recruit most heavily in schools, namely learning, attention, memory, decision-making, and social functioning, are both profoundly affected by and subsumed within the process of emotion (Immordino-Yang & Damasio, 2007).

9.3 Emotional Attachment, Cognition, and Behavior

Relationships between teachers and children have been a focus of educators' and psychologists' concerns for decades. In an academic setting, regulation of emotion is imperative because it can determine students' academic achievement and performance. Empathetic, supportive attachments, and relationships in school are essential to optimize brain development as "the attunement of emotional states is essential for the developing brain to acquire the capacity to organize itself more autonomously as the child matures" (Siegel, 2012). Research demonstrates that emotions fundamentally drive cognitive learning and to generate successful learning, educators need to engage the affective dimensions of pupils' minds (Immordino-Yang & Damasio, 2007). To be able to engage in learning a child needs to be able to take risks, learn new things, and face new challenges. A good learner needs to be able to manage frustration and anxiety, have good self-esteem, willing to take a risk and be able to ask for help when needed. Children who can regulate their own emotions and responses are more popular, have fewer behavioral problems, more emotionally stable, have a fewer infectious illnesses, and achieve more academically in schools (Graziano et al., 2007). A healthy educational environment is one where the interaction between the teacher, the pupil, and the learning task share a fluid dynamic, where the task reflects a teacher's awareness and understanding of pupil, and the pupil can seek reliable support when challenged by the task. Each related to the other in a way that fosters curiosity and supports the uncertainty that can be created by the challenges of "not knowing" which is at the heart of all learning (Geddes, 2006). Gradually, students experience a sense of their independent thoughts, ideas, and actions—an identity of their own, their capabilities. Henceforth, emotional attachment helps in enhancing students' experience of autonomy and independence and their possibilities of learning and growth (Geddes, 2006). In this context, Nagel once said (Nagel, 2009), "Emotional well-being must be a larger part of any learning, and by association, the educational agenda… Schools may be the optimum sites for buffering the impact of stress, building resilience and enhancing individual capacities for learning." In contrast, if a student experiences the world as unpredictable, irresponsive and/or hostile, this infuses a tremendous amount of energy in self-managing emotional arousal. A non-supportive attachment is associated with emotional and social incompetence, particularly in the areas of emotional understanding and regulated anger and anxiety (Mikulincer & Shaver, 2012). Furthermore, the perception of indifferent or aggressive teacher's behavior influences subsequent emotional responses and interpersonal behavior of students. For example, a student who experiences maltreated or feels threatened by a teacher's behavior often develops dislike, fear, or anxiety.

Research on emotional attachment gives importance to the relational rather than behavioral framework for the development of effective children's behavior (Bergin & Bergin, 2009; Cozolino, 2013; Riley, 2010). A behavioral framework that relies entirely on rewards and punishment to modify children's behavior is not necessarily the most effective model and does not always work. In contrast, the relational framework acknowledges that all behavior is a form of communication and adopts a "no-blame" ethos and cooperative policy. Relational frameworks include the strategy of emotional coaching, which refers to the useful tools for supporting children's well-being, learning, and behavior. Emotional coaching is all about helping to become aware of their emotions and to manage their feelings particularly during the instance of "misbehavior." It enables practitioners to create an ethos of positive learning behavior and to have the confidence to de-escalate situations when behavior is challenging.

9.4 Emotional Coaching Strategy for Promoting Sustainable Emotional and Behavioral Well-Being

Emotional coaching is a communication strategy that supports young people to self-regulate and manage their stress responses. The term was coined by American psychologist John Gottman and is gaining ground in the UK as an effective way to nurture mental health and well-being in education settings. Children and young people's behavior particularly in school continue to be a cause for concern in all sectors of society (Shaughnessy, 2012). Shaughnessy (2012) has highlighted that schools are nowadays flooded with policies, initiatives, and strategies for developing sustainable positive behavior and psychological well-being among students. However, oftentimes the most commonly used strategy constitutes "reasonable force" in terms of discipline and punishment to ensure "Good behavior in school" and to strengthen teacher's statutory authority to discipline pupils for their misbehavior. This appears to re-think the behaviorist approach which is based on the premise that behavior can be controlled and modified via reward and punishment techniques. Researchers (Davis, 2003; Shaughnessy, 2012) have shown that the quality of teacher–student relationships shapes the classroom experience and emotional needs of children which in turn influence students' social, emotional, and cognitive development. For example, supportive, responsive, and attuned teachers are more likely to diminish maladaptive behavior and stress. Therefore, the prime need is to shift attention from behaviorism and re-focus on the humanistic approach which acknowledges the complexity of children's behavior and focuses on internal factors, rather than external control.

Emotional coaching (EC) is referred to as being embedded within a bio-psycho-social model for the universal promotion of a person's well-being through philosophy and technique that can be used in the everyday situation (Gus et al., 2015). The EC philosophy embraces the principle that nurturing relationships acknowledge the legitimacy and acceptability of emotional experiences and provide opportunities to support children's development of resilience. Research shows that emotionally coached children are more emotionally stable, achieve more academically, are more popular, and have comparatively fewer behavioral problems. EC is conceptualized to support children to understand their emotions and why these occur, as well as how to manage their emotions more effectively (Gus et al., 2015). EC operated under the premise that emotions such as fear, joy, distress, surprise, and disgust are innate and universal (Gus et al., 2015; Rose et al., 2016). EC conceptualizes heightened states of emotion as opportunities for connection with another, wherein one person can "coach" the other through their emotional experiences, to support them in learning to express their feelings in a more adaptive manner (Rose et al., 2016). In this regard, the importance of teacher–pupil relationship, as a resource that has been suggested to influence children's academic and social development (Ahnert et al., 2013; Liberante, 2012).

9.5 Teacher–Student Attachment and Emotional Development

Theories of emotional development that focuses on child's capacity to regulate and express emotions separately from their contexts, neglect the vital contribution of dyadic and social interactions in development (Dunn, Brown, & Beardsall, 1991; Payne, 2003). The development of internal regulation and the modeling of normative behavior are associated with primary attachment figures, and teachers are one among them (Seth-Smith, 2006). Schools are inherently secondary caregiving environment and offer a fair opportunity for children's emotional development. As adults who form significant relationships with children, teachers can evoke attachment behavior, and there is evidence that teachers can become attachment figures, particularly for children with problematic attachment experiences. Sensitive teacher's responses to troubled students can facilitate better adaptations to everyday learning experiences (Bennathan & Boxall, 2000). Teachers, like parents, can shape a child's ability to regulate his or her emotions and to interpret other's emotional signals (Denham, 1998).

The influence of dyadic child–teacher interaction on emotional development is most powerfully explained within the framework of attachment theory. Bowlby's theory (Bowlby, 1969/1982, 1980) proposed that internal working models shape expectations of relationships through childhood and beyond. Categories of internal working models comprise secure, insecure, and disorganized (Bretherton & Mulholland, 1999; Main & Solomon, 1990). These underpin the "cognitive-affective systems" which underlie most areas of development, via their effects on a child's capacities to play and explore their levels of fearfulness and their interpersonal relationships (Mohanty & Suar, 2013a). In young children, the "attachment behavioral system" relates to biologically based exploratory and fear behavioral systems (Cassidy, 1999, p. 7). Distress activates the attachment system (the need to maintain proximity with a secure base, Bowlby, 1969/1982). Whereas a secure base is only intermittently present or is unavailable, heightened fearfulness, resulting from repeated exposure to anxiety-provoking experiences, activates the "defensive" cognitive and emotional structures which correspond with insecure/disorganized attachment representations. A range of risk and protective factors that operate in schools are associated with primary secure or insure internal working models. The sense of security is paramount to academic performance, in that when a child feels safe and able to face challenges, explore new arenas and therefore learn more (Piaget, 1955). The quality of children's relationships with teachers is particularly important in early childhood when children are likely to openly exhibit attachment related to behavior toward the teacher.

9.6 Emotional Competence, Emotional Intelligence, and Emotional Literacy

The development of emotional competence is related to the development of cognitive skills and it manifests differently in different ages. Notably, emotional competence does not develop in isolation from each other, and their progression is intimately tied to cognitive development. For example, insight into others' emotions grows in interaction with expanding awareness of one's own emotional experience. Some prominent emotional researchers have focused on the development of emotional intelligence which claimed by much social competence and academic attainment (Brenner & Salovey, 1997; Goleman, 1996). Research shows that "emotional intelligence[4] is associated with a wide range of positive outcomes among children and adolescents, including improving cognitive and social functioning, psychological well-being, and higher academic performance" (Brackett & Simmons, 2015, p. 24). Proponents of emotional intelligence advance a series of arguments in favor of developing "emotion literacy" through the use of specific curricula and the promotions of thoughtful interpersonal interactions in schools (Kam, Greenberg, & Kusche, 2004; Kusche, 2002; Sharp, 2001). In terms of the teaching task, there is sufficient evidence that shows children learn better when they are both cognitively and emotionally engaged in tasks (Goldstein, 1999; Siraj-Blatchford, Sylva, Muttock, Gilden, & Bell, 2002). Goleman (1996) and others (e.g., Mortimer, 2003; Sharp, 2001) argued that emotional literacy is an important factor in the improvement of overall learning standards and should be seen as a "core competence"

[4] John Mayer and Peter Salovey are the ones who introduced emotional intelligence in 1990. Daniel Goleman published a book in 1995 that popularized emotional intelligence.

underlying young people's ability to develop self-awareness, empathy, impulse control, self-discipline, and compassion. Promoting these qualities is claimed to raise school morale and decrease levels of aggression (Sharp, 2001).

9.7 Emotion and Attitude Development

Whether or not emotions are expressed explicitly through words and overt behaviors, they often manifest themselves in the form of implicit messages to which others consciously or subconsciously respond. Emotions manifest themselves in a positive or negative attitude toward the subject. Approaching and avoiding behaviors are based on these attitudes. An individual's emotional response is based, in part, upon the way he or she perceives implicit "information about feelings and like-dislike or attitudes" from others (Mehrabian, 1981, p. 3). Russell has proposed a three-factor model of human emotional response. According to this theory, all emotional states may be adequately described in terms of three independent dimensions: (a) pleasure–displeasure, (b) arousal–non-arousal, and (c) dominance–submissiveness. Each dimension is continuous and has within its range positive and negative values as well as a neutral point. Combinations of various values on each dimension characterize different emotions.

Many other researchers (Biggers, 1990; Vinson & Biggers, 1993) have supported Mehrabian and Russell's (1974) suggested three-factor emotional response theory. According to them (Vinson & Biggers, 1993), the above-mentioned three dimensions (pleasure, arousal, and dominance) combine to permit predictions of a higher-order construct called liking. The greater the degree of liking is felt by the person, the greater the likelihood of approach behaviors. The human emotional response thus permits predictions of approach or avoidance. The overall implications of the effect of emotional response upon behavior are that increased liking will result in greater approach behavior. Increased disliking will result in greater avoidance. In the context of a classroom, approach behavior could be operationalized to increase academic achievement. The following model describes the relationship between the predicted effect of teachers' behavior, the mediating presence of students' emotional response and their learning.

9.8 Social-Emotional Development and Cooperative Learning

Social-emotional development is defined as the acquisition of appropriate social skills needed to maintain appropriate peer and adult relationships (e.g., coping skills, listening, understanding non-verbal and verbal cues, and following directions) (Knitzer, 2003). Social-emotional learning is the process through which children and adults acquire and effectively apply the knowledge, attitudes, and skills necessary to understand and manage emotions, achieve positive goals, feel and show empathy for others, establish and maintain positive relationships, and make responsible decisions (Mohanty, & Suar, 2013b). Social-emotional learning is critical to learners' success. Many researchers have reported that teachers who show "warmth, caring, individual responsiveness to their students" have higher level engaged students and his/her students are usually enjoy learning, more hard-worker, and interact genuinely with their peers (Rimm-Kaufman, Baroody, Larsen, Curby, & Abry, 2015, p. 182). "When an instruction for social-emotional development is balanced with education for cognitive and physical growth, then we truly begin to educate the whole child" (Elmore & Zenus, 1994, p. 6). The importance of social-emotional development and its links to academic success starts from a very early age and has long-lasting effects. Educators, along with developmental psychologists, believe that it is essential to have a strong focus on emotional development programming for students, and it must begin as early as possible (Denham, Bassett, Zinsser, & Wyatt, 2014; Schonert-Reichl et al., 2015). There are a variety of studies spanning from preschool age to upper elementary that prove this connection to be true (Bavarian et al., 2013; Behforooz, Capece, & Vespo, 2006; Denham et al., 2014; Reid, Diperna, Missall, & Volpe, 2014; Schonert-Reichl et al., 2015; Segrott, Rothwell, & Thomas, 2013). To do this, today's school must provide social-emotional development and character because the bond made between students and teacher is the key in the development of the whole child (Durlak, Domitrovich, Weissberg, & Gullotta, 2015; Rimm-Kaufman et al., 2015; Schonert-Reichl et al., 2015; Segrott et al., 2013; Vespo, Capece, & Behforooz, 2006). However, the fact is that each year a student needs to create a new bond to a different teacher and so forth school bonding becomes crucial in school competence for every student in every passing year (Behfooroz et al., 2006; Murray & Greenberg, 2001). In this context, intervention programs can effectively provide opportunities for students to feel connections to their classrooms, which increase their academic success. The more students spent in a supportive and nurturing classroom, the more comfortable they felt to take an intelligent academic and social risk (Vespo et al., 2006). For this, the fundamental purpose of education is not merely to enrich students with the fullness of knowledge but also to establish the bond of love and friendship between man and man.

At an early age, children have very little understanding of their emotions in comparison to adults. Teachers through face-to-face interaction, sharing of knowledge and personal experiences, and encouraging co-operative learning[5] among students can promote this aspect of human psychology. Therefore, an interactive knowledge sharing education setting instigates emotional development in young minds. Understanding the emotion of others and one's own is necessary for building up a union with the surrounding. Cooperatively taught students tend to exhibit high-level

[5] Cooperative learning is an educational approach that aims to organize classroom activities into academic and social learning experiences.

reasoning and critical thinking skills, deeper understanding of learned material, greater time on task and less disruptive behavior, lower levels of anxiety and stress, greater intrinsic motivation to learn and achieve, greater ability to view situations from other' perspectives, more positive and supportive relationships with peers, more positive attitudes toward subject areas, and higher self-esteem (Felder & Brent, 2007). Henceforth, social-emotional development is one of the essential aspects of the creative process which includes the child's experience, expression, and management of emotions and the ability to establish positive and rewarding cooperative relationships with surroundings (Cohen, Onunaku, Clothier, & Poppe, 2005). In learning to recognize, label, manage, and communicate emotions and to perceive and attempt to understand the emotion of others, children build skills which essentially implies self-awareness, self-regulation, motivation, and empathy (Bhalla & Nauriyal, 2004; Kierstead, 1999) that connect them with family, peers, teachers, and the community. Thus, emotional bonding has triggered the process of unity with surroundings. In his view, this unity can only be achieved through education. Schools serve as the prime location for the promotion of emotional development (Tiwari & Srivastva, 2004) through everyday activities, interactions, and responses at school (Bloom, 1964; Hunt, 1961).

9.9 Emotional Labor and Education

The term emotional labor was first coined by sociologist Arlie Hochschild in her 1983 book *The Managed Heart*. Faking or suppression of true emotion and its display on demands of a job at the workplace is called emotional labor. At first glance, the skills of emotional labor might appear to be necessary only for the most rudimentary types of service work. However, recent research has documented that these skills are critical to success in a variety of professional and paraprofessional occupations, including medicine, law enforcement, teaching, administration, and emergency responses (Tracy, 2005). Emotional labor is one of the valued and demanding human capital in knowledge and service work, but unfortunately, the educators and students both do not recognize its significance in the twenty-first-century labor market (Lewis & Haviland-Jones, 2008). Psychologist Paul Ekman (Ekman & Friesen, 1975) added the complementary term "display rules" to refer to the visible manifestations of inner feelings or overt expressions that are normatively expected to be shown in a particular situation. Individuals begin to learn the expectations for their feelings and expressions of them as emotional work. Later, they learn to transfer these abilities as emotional labor or to comport themselves appropriately in the workplace. In this regard, it is believed that schools can play an important role in preparing students for future jobs by encouraging and deliberately strategies opportunities to practice the range and nuances of emotional labor through interactions with teachers and peers or role-play games.

On the other hand, teachers are expected to adjust and control their emotions and emotional displays everyday—whether it is with students, administrators, or parents

(Hargreaves, 1998; Schutz & Zembylas, 2009). For teachers, "emotions are intimately involved in virtually every aspect of the teaching and learning process and, therefore, an understanding of the nature of emotions within the school context is essential" (Schutz & Zembylas, 2009). A steady increase in studies regarding teachers' emotions reveals that emotions related to teaching might be associated with teachers' burnout, job satisfaction, health issues, and high rate of attrition (Schutz & Zembylas, 2009; Sutton, 2004). Where managing emotion is considered an integral part of a teacher's skillset, but for teachers, it is crucial to find appropriate strategies for emotion management (Fried, 2011). Some researchers claim that a teacher's emotional experiences are related to classroom management, discipline, and interactions with students, which would potentially lead to achieving teaching goals (Lee et al., 2014). Moreover, the task of grounding the emotional labor in a formal and systematic way requires clarification through education and the opportunity to combine a variety of theoretical models such as the sociological (Hochschild, 1983) and the phenomenological (Benner, 1994), to make emotional labor visible, valued and understandable. This will further help to minimize exploitation and exhaustion of emotional needs.

9.10 Conclusion

When parents have concerns about their children's overall development and education, teachers are the prime point-of-contact for information and assistance. A teacher plays a range of roles in children's development, including their academic achievement, also their psychological and behavioral health. As a result, focusing on children's mental health and the role of teachers is both globally relevant and necessary in the wider discourse on supporting healthy and educated children so that they can participate fully in society. In today's schools there is a great emphasis on standardized testing and teacher accountability, resulting in the emotions both of the students and teachers being pushed aside to the detriment of all (Brackett & Simmons, 2015). Knowing the sad reality at this time that delivery of the academic curriculum continues to be the sole driver of time and effort of every school, any amount of attention paid to developing social and emotional well-being will benefit the students at large is mostly ignored (Steinberg, 2015; Tomlinson, 2015). Although through the use of direct assessment and a reliable observation system, a teacher can understand the strengths and weaknesses related to the social-emotional development of a student. For this reason, explicit social emotional teaching and learning must be embedded in the curriculum for teachers' training. Adherence considering the teachers' needs is essential in providing healthy learning environments for all. When attention and care are given toward creating emotional intelligence in schools, less stress and burnout along with greater job satisfaction has been reported among educators (Brackett & Simmons, 2015; Collie, Shapka, & Perry, 2012). Teachers must be properly trained and prepared to support their students in their social-emotional development (Collie et al., 2012). This takes time, thought, and effort to carry out with fidelity. Every teacher has a different degree of experience and knowledge about child development, which could affect how the programs and strategies are carried out in the classroom (Reid et al., 2014). It is therefore important to consider the level of teacher's comfort and knowledge when implementing social-emotional development strategies (Collie et al., 2012). Most of the teachers training programs are focused on supporting teachers not only in delivering the course and curricula details and school policies but also include a detail discussion on how to better develop their social and emotional competence to promote their own healthy lifestyle and work environment (Segrott et al., 2013). As teachers need to be well equipped to promote social-emotional development to all learners, the teachers' training program implementers must take into consideration exactly what the teachers need and what areas require professional assistance (Rechtschaffen & Rechtschaffen, 2015).

References

Ahnert, L., Milatz, A., Kappler, G., Schneiderwind, J., & Fischer, R. (2013). The impact of teacher–child relationships on child cognitive performance as explored by a priming paradigm. *Developmental Psychology, 49*(3), 554.

Allen, G. (2011). *Early intervention: The next steps: An independent report to HM Government*. London, UK: The Cabinet Office.

Bavarian, N., Lewis, K. M., DuBois, D. L., Acock, A., Vuchinich, S., Silverthorn, N., & Flay, B. R. (2013). Using social-emotional and character development to improve academic outcomes: A matched-pair, cluster-randomized controlled trial in low-income, urban schools. *Journal of School Health, 83*(11), 771–779.

Behforooz, B., Capece, D. A., & Vespo, J. E. (2006). Effects of the nurturing curriculum on social, emotional, and academic behaviors in kindergarten classrooms. *Journal of Research in Childhood Education, 20*(40), 275–285.

Bennathan, M., & Boxall, M. (2000). *Effective interventions in primary school: Nurture groups*. London, UK: David Fulton.

Benner, P. (1994). Caring as a way of knowing and not knowing. In S. Phillips & P. Benner (Eds.), *Crisis of care* (pp. 42–62). Georgetown: Georgetown University Press.

Bergin, C., & Bergin, D. (2009). Attachment in the classroom. *Educational Psychology Review, 21*, 141–170.

Bhalla, S., & Nauriyal, D. K. (2004). Emotional intelligence: The emergency paradigm in personnel dynamics. *Psychological Studies, 49*(2), 97–106.

Biggers, T. (1990). Implicit communication theory.. Paper presented at the annual meeting of the International Communication Association, Miami, FL.

Bloom, B. S. (1964). *Stability and change in human characteristics*. New York, NY: John Wiley and Sons.

Bower, G. H., & Forgas, J. P. (2001). Mood and social memory. In J. Forgas (Ed.), *Handbook of affect and social cognition* (pp. 95–120). New York, NY: Psychology Press.

Bowlby, J. (1969/1982). *Attachment and loss* (Attachment) (Vol. 1). New York, NY: Basic Books.

Bowlby, J. (1980). Attachment and loss: Vol. 3: Loss. Hogarth Press and the Institute of Psycho-Analysis.

Bowlby, J. (1988a). *A secure base*. New York, NY: Basic Books.

Bowlby, J. (1988b). *A secure base: Parent-child attachment and healthy human development*. London, UK: Routledge.

Brackett, M., & Simmons, D. (2015). Emotions matter. *Educational Leadership, 73*(2), 22–27.

Brenner, E. M., & Salovey, P. (1997). Emotion regulation during childhood, developmental, interpersonal, and individual considerations. In P. Salovey & D. J. Sluyter (Eds.), *Emotional development and emotional intelligence: Educational implications* (pp. 168–193). New York, NY: Basic Books.

Bretherton, I., & Mulholland, K. A. (1999). Internal working models in attachment relationships: A construct revisited. In J. Casssidy & P. R. Shaver (Eds.), *Handbook of attachment. Theory, research and clinical applications*. New York, NY: The Guildford Press.

Cassidy, J. (1999). The nature of the child's ties. In J. Casssidy & P. R. Shaver (Eds.), *Handbook of attachment. Theory, research and clinical applications* (pp. 3–20). New York, NY: The Guildford Press.

Cohen, J., Onunaku, N., Clothier, S., & Poppe, J. (2005). *Helping young children succeed: Strategies to promote early childhood social and emotional development.* Washington, DC: National Conference of State Legislatures.

Collie, R. J., Shapka, J. D., & Perry, N. E. (2012). School climate and social-emotional learning: Predicting teacher stress, job satisfaction, and teaching efficacy. *Journal of Educational Psychology, 104*(4), 1189–1204.

Cozolino, L. (2013). *The social neuroscience of education: Optimizing attachment and learning in the classroom.* London, UK: Norton & Co.

Davis, H.A. (2003). Conceptualizing the role and influence of student-teacher relationships on children's social and cognitive development. *Educational Psychologist, 38*(4), 207–234.

Denham, S. A. (1998). *Emotional development in young children* (The Guilford Series on social and emotional development). New York, NY: Guilford Press.

Denham, S. A., Bassett, H. H., Zinsser, K., & Wyatt, T. M. (2014). How preschoolers' socialemotional learning predicts their early school success: Developing theory-promoting, competency-based assessments. *Infant and Child Development, 23*(4), 426–454.

Dunn, J., Brown, J., & Beardsall, L. (1991). Family talk about feeling states and children's later understanding of others' emotions. *Developmental Psychology, 27*(3), 448.

Durlak, J. A., Domitrovich, C. E., Weissberg, R. P., & Gullotta, T. P. (2015). *Handbook of social and emotional learning: Research and practice*. New York, NY: Guilford Press.

Ekman, P., & Friesen, W. (1975). *Unmasking the face*. Englewood Cliffs, NJ: Prentice-Hall.

Elmore, R. F., & Zenus, V. (1994). Enhancing social-emotional development of middle school gifted students. *Roeper Review, 16*(3), 182–185.

Felder, R. M., & Brent, R. (2007). Cooperative learning, Chapter-4. In P. A. Mabrouk (Ed.), *Active learning: Models from the analytical sciences* (ACS Symposium Series) (Vol. 970, pp. 34–53). Washington, DC: American Chemical Society.

Fried, L. (2011). Teaching teachers about emotion regulation in the classroom. *Australian Journal of Teacher Education, 36*(3), 117–127.

Geddes, H. (2006). *Attachment in the classroom: The links between children's early experience, emotional wellbeing and performance in school*. London, UK: Worth Publishing.

Goldstein, L. S. (1999). The relational zone: The role of caring relationships in the co-construction on mind. *American Educational Research Journal, 36*, 647–673.

Goleman, D. (1996). *Emotional intelligence: Why it can matter more than IQ*. London, UK: Bloomsbury Publishing.

Graziano, P. A., Reavis, R. D., Keane, S. P., & Calkins, S. D. (2007). The role of emotion regulation in children's early academic success. *Journal of School Psychology, 45*(1), 3–19.

Greenwald, A. G., Banaji, M. R., Rudman, L. A., Farnham, S. D., & Nosek, B. A. (2000). In J. P. Forgas (Ed.), *Feeling and thinking. The role of affect in social cognition* (pp. 308–331). Cambridge, UK: Cambridge University Press.

Gus, L., Rose, J., & Gilbert, L. (2015). Emotion coaching: A universal strategy for supporting and promoting sustainable emotional and behavioural well-being. *Educational & Child Psychology, 32*(1), 31–41.

Hargreaves, A. (1998). The emotional politics of teaching. *Teacher and Teacher Education, 14*(8), 835–854.

Hochschild, A. (1983). *The managed heart*. Berkeley: University of California Press.
Hunt, J. M. (1961). *Intelligence and experience*. New York, NY: The Ronald Press.
Immordino-Yang, M. H., & Damasio, A. (2007). We feel, therefore we learn: The relevance of affective and social neuroscience to education. *Mind, Brain, and Education, 1*(1), 3–10.
International Statistical Classification of Diseases and Related Health Problems. Tenth Revision. Vol. 1: Tabular list, 1992. Vol. 2: Instruction Manual, 1993. Vol. 3: Index. Geneva, World Health Organization.
Kam, C., Greenberg, M. T., & Kusche, C. A. (2004). Sustained effects of the PATHS curriculum on the social and psychological adjustment of children in special education. *Journal of Emotional and Behavioral Disorders, 12*, 66–78.
Kierstead, J. (1999). *Human resource management trends and issues: Emotional intelligence (EI) in the workplace*. Research Directorate, Policy Research and Communications Branch, Public Service Commission Branch, Public Service Commission of Canada.
Knitzer, J. (2003). *Social and emotional development in young low-income children: What research tells us and why it matters for early school success*. Paper presented at National Center for Children in Poverty, New York, NY.
Kusche, C. A. (2002). Psychoanalysis as prevention: Using PATHS to enhance ego development, object relationships & cortical integration in children. *Journal of Applied Psychoanalytic Studies, 4*, 283–300.
Lee, M., Schutz, P. A., Taxer, J., Pekrun, R., Meier, E., & Xie, X. (2014). *Teachers' emotions, emotion regulation, and classroom management efficacy*. Paper presented at the Annual Meeting of the American Educational Research Association, Philadelphia, USA.
Lewis, M., & Haviland-Jones, J. M. (2008). *Handbook of emotions* (3rd ed.). New York, NY: Guilford Press.
Liberante, L. (2012). The importance of teacher-student relationships, as explored through the lens of the NSW Quality Teaching Model. *Journal of Student Engagement: Education Matters, 2*(1), 2–9.
Linnenbrink, E. A. (2007). The role of affect in student learning: A multi-dimensional approach to considering the interaction of affect, motivation, and engagement. In P. A. Schutz & R. Pekrun (Eds.), *Emotion in education* (pp. 107–124). San Diego, CA: Elsevier Academic Press.
Main, M., & Solomon, J. (1990). Procedures for identifying infants as disorganised/disorientated during the Ainsworth strange situation. In M. T. Greenberg, D. Cicchetti, & E. M. Cummings (Eds.), *Attachment in the preschool years* (pp. 121–160). Chicago, IL: University of Chicago Press.
Medrano, L. A., Flores, K. P. E., Ríos, M., Curarello, A., & González, J. (2015). *Adaptación de la Escala de AfectoPositivo y Negativo (PANAS) a la población de estudiantesuniversitarios de Córdoba*. Anuarios de Investigaciones de la Facultad de Psicología. Article in Edition.
Mega, C., Ronconi, L., & De Beni, R. (2014). What makes a good student? How emotions, self-regulated learning, and motivation contribute to academic achievement. *Journal of Educational Psychology, 106*(1), 121–131.
Mehrabian, A., & Russell, J. A. (1974). The basic emotional impact of environments. *Perceptual and Motor Skills, 38*(1), 283–301.
Mehrabian, A. (1981). Silent messages: implicit communication of emotions and attitudes, Albert Mehrabian, 2. edn. Wadsworth, Belmont, Calif.
Mikulincer, M., & Shaver, P. R. (2012). An attachment perspective on psychopathology. *World Psychiatry, 11*(1), 11–15.
Mohanty, S. N., & Suar, D. (2013a). Influence of mood states, group discussion, and interpersonal comparison on change in decision-making and information processing. *International Journal of Decision Sciences, Risk and Management, 5*(2), 101–123.
Mohanty, S. N., & Suar, D. (2013b). Decision-making in positive and negative prospects: Influence of certainty and affectivity. *International Journal of Advances in Psychology, 2*(1), 19–28.
Mohanty, S. N., & Suar, D. (2014). Decision making under uncertainty and information processing in positive and negative mood states. *Psychological Reports, 115*(1), 91–105.
Mortimer, H. (2003). *Emotional literacy and mental health in the early years*. Lichfield, UK: QEd.

Murray, C., & Greenberg, M. T. (2001). Relationships with teachers and bonds with school: Social emotional adjustment correlates for children with and without disabilities. *Psychology in the Schools, 38*(1), 25–41.
Nagel, M. (2009). Mind the mind: Understanding the links between stress, emotional well-being and learning in educational context. *International Journal of Learning, 16*(2), 33–42.
Payne, A. L. (2003). Measuring preschool children's perceptions of teacher-child relationship quality: Developing and testing the "feelings about my school and teachers' instrument". *Dissertation-Abstracts-International-Section A: Humanities-and-social-sciences, 64*(2-A), 391.
Piaget, J. P. (1955). *The language and thought of the child.* New York, NY: Meridian Books.
Ranellucci, J., Hall, N. C., & Goetz, T. (2015). Achievement goals, emotions, learning, and performance: A process model. *Motivation Science, 1*, 98–120.
Rechtschaffen, D., & Rechtschaffen, T. (2015). The 5 literacies of mindful learning. *Educational Leadership, 73*(2), 58–62.
Reid, E. E., Diperna, J. C., Missall, K., & Volpe, R. J. (2014). Reliability and structural validity of the teacher rating scales of early academic competence. *Psychology in the Schools, 51*(6), 535–553.
Riley, P. (2010). *Attachment theory and teacher student relationships*. London, UK: Routledge.
Rimm-Kaufman, S. E., Baroody, A. E., Larsen, R. A., Curby, T. W., & Abry, T. (2015). To what extent do teacher-student interaction quality and student gender contribute to fifth graders' engagement in mathematics learning*? Journal of Educational Psychology, 107*(1), 170–185.
Rose, J., Temple, S., McGuire-Snieckus, R., Wood, F., Vatmanides, O., & Parker, R. (2016). Somerset emotion coaching project evaluation report. Institute for Education, Bath Spa University.
Schonert-Reichl, K. A., Oberle, E., Lawlor, M. S., Abbott, D., Thomson, K., Oberlander, T. F., & Diamond, A. (2015). Enhancing cognitive and social–emotional development through a simple-to-administer mindfulness-based school program for elementary school children: A randomized controlled trial. *Developmental Psychology, 51*(1), 52.
Schutz, P. A., & Zembylas, M. (Eds.). (2009). *Advances in teacher emotion research: The impact on teachers' lives*. New York, NY: Springer.
Segrott, J., Rothwell, H., & Thomas, M. (2013). Creating safe places: An exploratory evaluation of a school-based emotional support service. *Pastoral Care in Education, 31*(3), 211–228.
Seth-Smith, F. (2006). *How do teachers influence children's emotional development?* Doctoral dissertation, University of London.
Sharp, P. (2001). *Nurturing emotional literacy. A practical guide for teachers, parents and those in the caring professions*. London, UK: David Fulton.
Siegel, D. (2012). *The developing mind: How relationships and the brain interact to shape who we are*. New York, NY: Guildford Press.
Siraj-Blatchford, I., Sylva, K., Muttock, S. Gilden, R., & Bell, D. (2002). *Researching effective pedagogy in the early years.* Research Report No.: RR356. Nottingham, UK: DfES.
Shaughnessy, J. (2012). The challenge for English schools in responding to current debates on behaviour and violence. *Pastoral Care in Education, 30*(2), 87–97.
Steinberg, L. (2015). How self-control drives student achievement. *Educational Leadership, 73*(2), 28–32.
Sutton, R. E. (2004). Emotional regulation goals and strategies of teachers. *Social Psychology of Education, 7*(4), 379–398.
Tiwari, P. S. N., & Srivastva, N. (2004). Schooling and development of emotional intelligence. *Psychological Studies, 49*(2–3), 151–154.
Tomlinson, C. A. (2015). Teaching for excellence in academically diverse classrooms. *Society, 52*(3), 203–209.
Tracy, S. J. (2005). Locking up emotion: Moving beyond dissonance for understanding emotion labor discomfort. *Communication Monographs, 72*(3), 261–283.
United Nations General Assembly. (2015). *Resolution adopted by the General Assembly on 25 September 2015*. Washington, DC: United Nations.

Vespo, J. E., Capece, D., & Behforooz, B. (2006). Effects of the nurturing curriculum on social, emotional, and academic behaviors in kindergarten classrooms. *Journal of Research in Childhood Education, 20*(4), 275–285.

Vinson, L., & Biggers, 3. T. (1993). Emotional response as a predictor of compliancegaining message selection. *The Southern Communication Journal, 58*, 192–206.

Chapter 10
Emotional Level Classification and Prediction of Tweets in Twitter

E. Rajesh Kumar, A. K. V. S. N. Rama Rao, and Soumya Ranjan Nayak

Contents

10.1 Introduction

At present, people are influenced by social media. It has become more significant part of human life, and Twitter is one of well-known among them due to its growth and increase in Twitter users. Approximately there are 145 million daily and 332 billion monthly active Twitter users (Millions, 2008) excluding company agents, celebrities (VIPs), and politicians. Fundamentally, Twitter supports only short messages called tweets with 140 characters in length. In spite of the fact that tweets are short, yet they are significant for expressing and sharing emotions. Using these short tweets, any individual can express their feelings as positive or negative regarding any topic they face in daily life. As feeling or emotions impact everything in daily life, investigators handle this as a challenge to recognize and experiment emotion. Implicit Feedback generated using Artificial Intelligent (AI) System from individuals helps in adjusting the customer's experience (e.g., communication with virtual

E. Rajesh Kumar · A. K. V. S. N. Rama Rao
Department of Computer Science and Engineering, Koneru Lakshmaiah Education Foundation, Vaddeswaram, AP, India

S. R. Nayak (✉)
Amity School of Engineering and Technology, Amity University Noida, Uttar Pradesh, India

S. N. Mohanty (ed.), *Emotion and Information Processing*,
https://doi.org/10.1007/978-3-030-48849-9_10

operators) (Rahid, Ali Sha, Carsten, Mohsin, & Wahid, 2019; Sebe, Cohen, & Huang, 2005). Detection of intimidation on social media (Dinakar, Jones, Havasi, Lieberman, & Picard, 2012).

To observe the feelings of an individual, different devices like skin conductance, sensors, electromyography, speech, body posture, and even text are used (Sebe et al., 2005; Wang, Chen, Thirunarayan, & Sheth, 2012). Sensors used for such prediction is more expensive and can be operable only trained people. To bring feeling recognition to each individual and request for emotion, feeling recognition should be enabled empowered on devices that can be accessible to overall population. The ideal examples of such gadgets are smartphones and PCs that allow any individuals to send message via any social media sites.

Text-based platforms widely uses Facebook, Twitter, and SMS for communication. The main objective of this chapter is to improve emotion detection on text data. Such detection is challenging for two reasons: (1) keyword-based detection was easy to identify emotion, for example, "Hate" fit in to different emotion; embarrassment and anger, (2) feelings are hidden in words without any reference to text, for example, "panic" and "scare" (Wang et al., 2012). So, the real-time emotion detection has to function regularly to add feelings detection to real-time services like Twitter. Such detection is done in proposed system by finding the probability of each term in a document and higher probability term, and by smoothing the data, the emotion level can be calculated.

10.2 Related Works

Sentiment analysis was proposed by various researchers with different techniques. Some are based on probabilistic features, dictionary (lexicon) features, Artificial Intelligence techniques, and even with the combination of two or more techniques. Agarwal, Boyi, Vovsha, Owen, and Rebecca (2011) constructed two different classification models, a three-way model that can classify text into classes like positive, negative, and neutral and a two-way model to classify tweets into positive and negative classes, for analysis author used three models; tree kernel, feature model, and unigram model with annotated data consisting of positive, negative, and junk or neutral polarities. All tweets exhibiting junk polarities are eliminated. For polarity scoring, earlier they used "Dictionary of Affect in Language (DAL)" and proposed with WordNet. The author used Support Vector Machine (SVM) by considering the word as negative with polarity less than 0.5, positive for words with polarity greater than 0.8, and remaining polarity ranges of data as neutral and obtained an average result. Pak and Paroubek (2010) used a bulk of 300,000 Twitter data partitioned data into three sets, namely positive, negative, and no emotions. They (Pak & Paroubek, 2010) analyzed the data with positive and negative emotions by training the classifier using trigram, bigram, and unigram model. During filtering process data preprocessing is done by removing URLs, special symbols, stopwords, and by forming a bag of words author proposed multinomial Naive Bayes method and SVM and obtained a better result with Naive Bayes model.

Pang, Lee, and Vaithyanathan (2002) examined the sentiment utilizing machine learning method to classify the data into positive and negative polarities. The classification is done based on SVM, NB, and maximum entropy technique that attained a better result by using different feature for unigram data and not based on feature frequency. Alec, Bhayani, and Huang (2009) used only tweets with positive and negative polarity and eliminated neutral tweets in training and test data. In classification model without feature extraction, Alec et al. (2009) proposed SVM and maximum entropy and NB by clearing all the emoticons from the training model and produced low accuracy by using maximum entropy and SVM model when compared to NB technique. For better efficiency, Alec et al. (2009) used unigram, bigram, unigram and bigram feature, feature reduction by replacing tweets containing URLs ("https://tinyurl.com/cvvg9") with token "URL" and "USERNAME" with symbol "@" added to prefix of every class and tested with different classification models like NB, SVM, and key-based techniques, these models had produced a better efficiency with unigram and bigram model. Han and Kamber (2006) mainly focused on hashtags and emoticons and classified them into six emotions. SVM via LIBSVM with unigram and linear kernel feature were examined with three different analysis: (1) emotion perception, (2) emotion recognition, and (3) manually annotated labeling obtained three different results varied from 50% to 80% accuracy.

10.3 System Architecture

See Fig. 10.1.

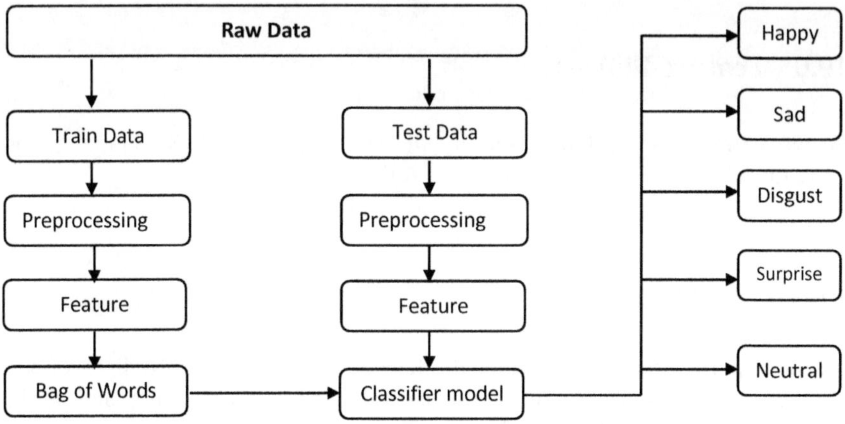

Fig. 10.1 Proposed model

10.4 Data Collection

Nowadays, topic related to sentiment analysis is very popular in research. There remains many structured data that are available online. Among them, Twitter APIs can also be utilized to access the Twitter tweets. But, dataset for emotional analysis does not exist online. The data are collected from the site sentiment140 that provides labeled quantity of data over 1,400,000 tweets of data with three different polarities (positive, negative, and neutral), and tweets are labeled manually. From the labeled tweets, some data may contain hashtags, symbols, and tweets with different language, all these data are eliminated in preprocessing level as explained in next paragraph, and data without emotional representation is also eliminated.

10.5 Data Preprocessing

All labeled data are preprocessed because of noisy data that are either long or short in length. The preprocessing steps are as follows.

- Acronym words are collected from online that are frequently used in social media like Twitter (Sebe et al., 2005). A separate dictionary is created with acronyms along with its English abbreviation. Example: omg: oh my god, why: y.
- Tweets with repeated characters are converted to normal characters. Example: "cooooool": "cool," "happyyyyy": "happy."
- Stemming and lemmatization are used to recognize word of base forms from reducing inflected form. For example, "car," "car's," "cars" are reduced to "car."
- All sentences with hashtags(#) symbol, '@' symbol, username, stopwords, and punctuation symbol are removed that do not represents any emotion to the sentence.

10.6 Feature Extraction

Feature extractions are done based on unigram model with POS tagging and unigram model. The extraction process is as follows.

10.6.1 Unigram

It is an important model for sentiment-related feature extraction. To discover probabilities for a series of terms, chain rule is used to break down the probabilities into successive events for all series of events. The formula use for chain rule is

$$P(x_1, x_2, \ldots, x_n) = P(x_1)P(x_2 \mid x_1)P(x_3 \mid, x_2 \mid, x_1)\ldots P(x_n \mid x_{n-1}, \ldots x_1) \quad (10.1)$$

where, "x_1, x_2,..., x_n" denotes indexes of random variables and "P" denotes probability.

This model is also a simplest language method that never consider training context and defines each term individually by the following formula:

$$P_{\text{unit}}(x_1, x_2, \ldots, x_n) = P(x_1)P(x_2)\ldots P(x_n) \quad (10.2)$$

All words hit and depend on their own due to its probability; only single finite state unit is used as automata.

10.6.2 Unigram with POS Tagging

In this model, every word is annotated with POS tagging. Different POS is selected and used for feature extraction; by using these tagging, it helps in separation of meaningful words from Twitter tweets by semantics and language philosophy (compositionality principle). All emotional words are termed as emotion rule, and the remaining non-emotional texts are eliminated. POS tagging used are as follows in Table 10.1.

10.6.3 Bab of Words (BOW)

It is used to find the frequency of a word as a feature in the training the model to tokenize the data into unigram for every class. All the text is decomposed into a single unit to store the term frequency (TF).

Table 10.1 POS tagging used

VBP—"verb, singular present, non-third person"	**RB**—"adverb"
VB—"verb, base form"	**RBS**—"adverb, superlative"
VBD—"verb, past tense"	**JJR**—"adjective, comparative"
VBN—"verb, past participle"	**JJ**—"adjective"
VBG—"verb, present participle"	**JJS**—"adjective, superlative"
UH—"interjection"	**NNP**—"proper noun, singular"
VBZ—"verb, third person singular"	**NN**—"noun, singular"
NNPS—"proper noun"	**NNS**—"proper noun, plural"

10.7 Proposed Classifier

10.7.1 Disadvantage of Existing Model

During emotion prediction, absence of words or if the frequency of words appeared with some probability in test data but not in training data, the NB model considers the data as zero frequency. But this drawback can be overcome in proposed model by using smoothing method.

10.7.2 Proposed Model

In proposed model, the main objective is to find the best class for text classification. So, multinomial NB classification is used; it is a type of NB classifier used for text classification (Han & Kamber, 2006). All the data "*d*" and class "*c*" the NB and the probability can be calculated by:

$$P(c \mid d) \alpha P(c) \prod_{1<=k<=nd} P(t_k \mid c) \tag{10.3}$$

where $P(t_k|c)$ denotes the conditional state probability of the class "*c*." So that, it highlights the amount of term available in class. "n_d" denotes the number of tokens in the document "*d*." "*P*(*c*)" denotes probability of "*d*" occurring in the "*c*." If the term in document provides less evidence compared to another class. The probability with higher term is chosen. Here (t_1, t_2, t_3,…t_n) are tokens in the document "*d*." In NB model, the top class is "maximum a posterior (MAP)" (Cmap) class:

$$\text{Cmap} = \arg\max_{c \in C} \hat{P}(c \mid d) = \arg\max_{c \in C} \hat{P}(c) \prod_{1<=k<n_d} \hat{P}(t_k \mid c) \tag{10.4}$$

Since the value of *P*(*C*) and $P(t_k|c)$ is unknown, in Eq. (10.4) "*P*" is denoted as " $\hat{P}$ " and $P(t_k|c)$ is multiplied for each $1 < =k < = n_d$ that may lead to decimal value (underflow). So, logarithmic is performed instead of multiplication. The class "*c*" with higher probability value is more feasible; log(*ab*) = log(*a*) + log(*b*) and it is monotonic. The proposed can reach a maximum:

$$\text{Cmap} = \arg_{c \in C} \max \left[\log \hat{P}(c) + \sum_{1<=k<=n_d} \log \hat{P}(t_k \mid c) \right] \tag{10.5}$$

In Eq. (10.5) "log $\hat{P}$ '(*C*)" indicates frequency, "log $\hat{P}$ ($t_k|c$)" indicates weight and the level of t_k of class "*c*." Classes with more frequency are correct when compared to nonfrequent class. The frequency and weight of a class are calculated by the formula:

$$P(C) = \frac{N_c}{N'} \tag{10.6}$$

where "N_c" is the number of contents or document and "N" indicates the total document in class "c." For probability of term "t" can be calculated by:

$$P(t \mid c) = \frac{T_{ct}}{\sum t \in vT_{ct'}} \tag{10.7}$$

In Eq. (10.7) "T_{ct}" denotes the occurrence of words in training document for class "c" associated duplicate occurrence of term. The maximum likelihood estimate (MLE) can be calculated for multiple class for a word. It has a drawback. For example, if a word "HELLO" occurred only in document name SAMPE, then MLE for different class will be "0."

$$\hat{P}(\text{HELLO|SAMPLE}) = 0 \tag{10.8}$$

Finally, the higher frequency word is indicated with "1" and low frequency is denoted as "0" for exploring emotion. But the "0" frequency data is avoided by smoothing process by adding an extra count for words with no value or words with different count.

10.8 Result and Analysis

The emotion level of the tweets can be explored using a four-way model by using surprise, happy, sad, and disgust emotion, five-way model by using neutral polarity associated with 4-way model. In the proposed model multinomial, NB classifier is used by extracting unigram and unigram with POS tagging features. Table 10.2 shows that the accuracy level of the four-way classification model achieved 82.25% better accuracy with unigram features when compared to unigram with POS tagging and F1-score of 95% for the emotion surprise. Table 10.3 shows the accuracy level for five-way classification model with different features, additional emotion "neutral" is included in this model and unigram feature produced 69.4% better accuracy compared to unigram with POS tagging. Table 10.4 shows the accuracy level of comparison of the proposed model with two different models, comparing the results with two different classification methods. The four-way model has achieved 82.25% better accuracy when compared to other systems. Alec et al. (2009) have produced accuracy with slight variation with the proposed method.

Table 10.2 Accuracy of four-way classification model

	Emotion				
Features used	Surprise (%)	Happy (%)	Sad (%)	Disgust (%)	Accuracy (%)
Unigram	91	81	85	72	82.2
Unigram with POS tagging	95	78	81	69	81

Table 10.3 Accuracy of five-way classification model

Features used	Emotion Surprise (%)	Happy (%)	Sad (%)	Disgust (%)	Neutral (%)	Accuracy (%)
Unigram	91	81	85	72	18	69.4
Unigram with POS tagging	95	78	81	69	14	67.3

Table 10.4 Accuracy comparison table

	Proposed system Accuracy		Accuracy of Agarwal et al. (2011)		Accuracy of Alec et al. (2009)
Feature used	Four-way model (%)	Five-way model (%)	Two-way model (%)	Three-way model (%)	NB with binary classification (%)
Unigram	82.25	69	71.35	56.58	81.3
Unigram with POS tagging	81	67.4	–	–	79.9

10.9 Conclusion

People prefer different online medium for communication especially microblogging (Tumblelogs) site that helps people to transfer images, videos, and sentences as a message. It has an excessive deal on social, personal, politics, and business. So, opinion mining or sentiment analysis of these Tumblelogs has unlimited mean due to more investigation on this site. In the proposed method, a new framework to analyze the emotions for Twitter tweets is built. The idea is to deal only with text contents that can be small sentence or even a big sentence. By training the data before test data, feature extraction, and applying the classification model by further smoothing of every class for unigram model has produced better result compared to binary and other models. The four-way classification method has achieved 82.25% accuracy compared to the existing system, two-way, three-way, and five-way classification, thus by providing a scalable method for calculating emotion level in Twitter data.

References

Agarwal, A., Boyi, X., Vovsha, I., Owen, R., & Rebecca, P. (2011). Sentiment analysis of twitter data. In *ACM Proceedings of the Workshop on Languages in Social Media* (pp. 30–38).

Alec, G., Bhayani, R., & Huang, L. (2009). Twitter sentiment classification using distant supervision. In *Proceedings of Empirical Methods in Natural Language Processing* (pp. 79–86).

Dinakar, K., Jones, B., Havasi, C., Lieberman, H., & Picard, R. (2012). Commonsense reasoning for detection, prevention, and mitigation of cyber bullying. *ACM Transactions on Interactive Intelligent Systems, 2*(3), 1.

Han, J., & Kamber, M. (2006). *Data mining* (pp. 311–312). Amsterdam, Netherlands: Elsevier.

Millions, N. (2008). *Twitter: Number of active users 2010–2018|Statista*. Statista, 2018 [Online]. Retrieved from http://www.statista.com/statistics/282087/number-of-monthlyactive-twitter-users

Pak, A., & Paroubek, P. (2010). Twitter as a corpus for sentiment analysis and opinion mining. In *International Conference on Language Resources and Evaluation* (pp. 1320–1326).

Pang, B., Lee, L., & Vaithyanathan, S. (2002). Sentiment classification using machine learning techniques. In *Proceedings of the ACL-02 Conference on Empirical Methods in Natural Language Processing* (Vol. 10, pp. 79–86). Association for Computational Linguistics.

Rahid, K., Ali Sha, M., Carsten, M., Mohsin, M., & Wahid, A. (2019). Emotion classification and crowd source sensing; A lexicon based approach. *IEEE Access, 7*, 27124–27135.

Sebe, N., Cohen, I., & Huang, T. S. (2005). Multimodal emotion recognition. In *Handbook of pattern recognition and computer vision* (pp. 387–409). Singapore: World Scientific.

Wang, W., Chen, L., Thirunarayan, K., & Sheth, A. (2012). Harnessing Twitter big data for automatic emotion identification. In *IEEE International Conference on Social Computing* (pp. 587–592).

Chapter 11
Remaining Useful Life as a Cognitive Tool in the Domain of Manufacturing

Ahin Banerjee, Sanjay K. Gupta, and Debanshee Datta

Contents

11.1 Introduction

Decision-making in expert systems unlike cognitive process selects the best from a set of alternatives (Y. Wang & Ruhe, 2007). The need for behavioural scientists in the area of cognitive science is to study the process and thereby examine the need, with the focus to transform information. Advancement and availability of big data are seen to be continually increasing in the network repositories. Real-time data processing and analysis are not yet conducted, and hence, control over such expanse in the volume of data by the automotive companies still remains insignificant. A promising performance improvement with big data analytics can be seen in the near future for data-driven industrial context. Cognitive action also known to be the highest form of intelligence prioritizes learning and decision-making from the monitored industrial data towards building the logic and planning via expert system. Performance and availability for such systems cater the need for the development of the prognostic process besides a decrease in the total cost of their assignments. Selection strategy for the type of maintenance on the basis of optimized cost is the area of challenge faced by industry experts. Decision-making process can be further classified as deterministic and stochastic based (Shmelova, Sikirda, Scarponi, & Chialastri, 2018). For an industry, deterministic decision-making is not widely

A. Banerjee (✉) · S. K. Gupta · D. Datta
Indian Institute of Technology (BHU), Varanasi, Uttar Pradesh, India
e-mail: skgupta.civ@iitbhu.ac.in

S. N. Mohanty (ed.), *Emotion and Information Processing*,
https://doi.org/10.1007/978-3-030-48849-9_11

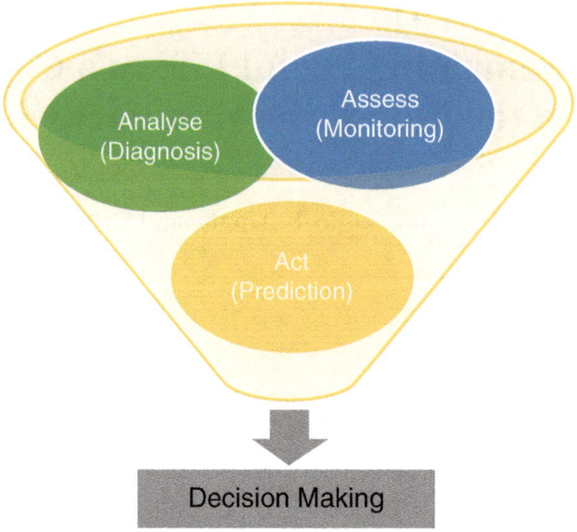

Fig. 11.1 Steps towards maintenance decision-making

acceptable because deterministic decisions are very rigid. To mitigate the risk factors the dynamic parameter 'safety' becomes a prime concern. The stochastic decision-making being evolutionary in nature on the other hand providing a narrow band of output values covers the underlying risk and uncertainty in the results, thereby reducing the amount of risk (Z. Wang, Busemeyer, Atmanspacher, & Pothos, 2013). Such evolutionary decision-making models is very much applicable towards industrial process are often reduces the problems faced by maintenance managers. To fulfil the above requirements in accordance to the organizational standards in selection of a suitable maintenance strategy besides supporting maintenance managers to '*Asses, Analyse and Act*' (Fig. 11.1) for their organization, a very convenient framework is discussed in this chapter.

This chapter deals with RUL-built maintenance policy for a deteriorating system whose condition is periodically monitored to build a human–machine interaction. Applications of artificial cognitive systems includes industrial robots and automated guided vehicles in space explorations to autonomous vehicles around us. In an OEM industry, system has to undergo quality assurance testing. This can be done by means of monitoring the health of the system through prognosis which is one application of PHM. Since, safety is a dynamic feature for any industry, RUL as an indicator of the health of the system is well formulated for maintenance-based decision-making in contrast to automotive domain.

The auto industry today uses more than half a ton of sensors implanted to the car body for data retrieval. The challenge is to aggregate and analyse such data in order to obtain useful insights. One such smart technology by means of evolutionary techniques for information processing with the use of data is currently enabled and are likely to be embedded into the physical assets. The use of prognostics in engineering advancement can be seen as a tool for providing information for stochastic decision-making and monitoring the equipment's state of health. Based on the monitoring information, an optimal predictive maintenance policy keeping in mind the budget and resources constraints must be planned. The current study attempts in developing one such efficient indicator a.k.a. RUL that will help towards inspection and maintenance decision-making. Section 2 features the past literary works and has been divided into three subsequent sections. Section 2.1 debates about the philosophy of decision-making with a weight towards cognition. Section 2.2 narrates the widely accepted maintenance-based monitoring techniques out of which the predictive maintenance is of choice in the current study Sect. 2.3 anecdotes the different modelling techniques and algorithm used for parameter estimation and RUL calculation. The possibility for RUL-based maintenance decision-making is yet to receive much attention and thereby providing a scope for the current study. Approach towards implementation is discussed in Sect. 3 where it provides a brief explanation about the methodology used and the results obtained. Conclusion to the study is attempted in Sect. 11.4.

11.2 Related Past Literary Works

11.2.1 Philosophy of Decision-Making with a Weight Towards Cognition

Prediction problems in engineering make use of the existing quantities of information (Shadlen and Kiani 2013) (S. N. Mohanty & Suar 2014) to predict the future outcomes from an organization's decision-making viewpoint. Critical review in medical decision-making (Patel, Kaufman, & Arocha, 2002) (S. N. Mohanty & Suar 2013a) was much acclaimed with the integration of problem-solving and decision-making in a broader perspective and valid foundation as limited to classical techniques. Evolution of Bayesian methodology for assessment of dynamical systems is seen as an advancement to describe the systems behaviour in terms of information flow. Wikipedia explains cognition as an "information processing" with respect to an individual's (machine) physical functions. This advancement in the process of learning besides decision-making is what it is today machine learning. With the rise in artificial intelligence technologies and embedded systems, the fault/injury to a machine needs to be featured as early as possible. Conventional cognitive models assume definite psychological state (S. N. Mohanty & Suar 2013b) at each moment; however, the true state remains unknown to an expert, thereby assigning some probability value (Z. Wang et al., 2013). Current study uses PF as an advanced (stochastic) technique to overcome this challenging work by providing an early signal to the user via state space model approach until a decision is reached.

11.2.2 Maintenance-Based Monitoring

For any industry to meet the quality standards, it follows four types of maintenance policies (Pham, 1994) that were formulated. First one relates to corrective maintenance, i.e., maintenance after failure of the asset besides it leverages for low value assets and challenging in case of random failures; the second follows the preventive maintenance, i.e., time-based maintenance (periodicity in maintenance between the failures) which applies to critical assets; third arises the preventive condition-based maintenance (CBM), here it uses statistical process control techniques for monitoring under regular interval as soon as the indicators cross the threshold alarm is triggered; and finally the fourth is the predictive maintenance which is an extension of the CBM that solves by forecasting the state or degradation level of the asset. The chapter dictates the use of predictive maintenance approach towards estimation of RUL. The RUL thus obtained will be used as a CI for decision-making purpose by the industry experts with economical conclusions.

11.2.3 Modelling Techniques and Algorithm Used for RUL Estimation

RUL estimation models based on directly observed state processes uses the condition monitored data and a known threshold value to calculate the RUL. Both the model and the threshold values are required for RUL estimation in the absence of historical data. Out of the two approaches towards RUL modelling for directly observed state process, Markovian-based discrete time space is followed in the present study. The predicted state then helps in estimating the RUL as a CI for the condition monitored data. The estimated RUL then acts as an alternative to the expert system which then takes the cognitive action to augment the safety. For a product manufacturing industry, there exists a wide variety of machines that can be subject to a class of failures. The inability to perform the intended function is defined as a failure. Failure originates from a faulty state (that may last either for a short or a longer period of time) likewise during conditions while the machinery is running or in a standby. The onset of incipient to degradation is also said to be a failure where the machine works but beyond undesired inputs. Condition monitoring of such machines may reveal early failures that helps in asset management. But based on gathered information and correlations, need for a universally accepted approach for finding the best estimate needs to be proposed, which however does not exist (Si, Wang, Hu, & Zhou, 2011). A practical approach towards appropriate prognostic algorithm selection is attempted using evolutionary-based stochastic approach (Banjevic, 2009), thereby formalizing the uncertainty, measurements sensitivity and the order of measurements (Moore, 2002). Fair reviews from data-driven and

Table 11.1 Two types of prognostic approaches for RUL finding

Model-based	Physical models/White box modelling/First principles modelling
Data-driven	Degradation based models/Stochastic models/Black box modelling

physics-based prognostics methods help in appropriate RUL selection. Table 11.1 categorically explains the use of two such widely known techniques.

Physics-based models are used to calculate the RUL of the system are basically information specific (e.g., material properties), besides they work well when along with the environmental and parameter data. The underlying dynamics limits the use of such models only to specialized class of researchers. Data-driven modelling use information from measured data to identify the properties of damage/degradation state and use it as a prior to update the posterior/future states. The presence of monotonic trend in the data features enables the handler to use this approach complex free. The data driven approach is further divided into two classes: a.k.a., artificial intelligence-based approach and the statistical models (An, Kim, & Choi, 2015). The subsequent section summarizes few industrial research works based on the above two classes.

A number of information processing techniques (Jardine, Lin, & Banjevic, 2006) from rotating machinery are reviewed that uses temperature, vibration and crack length as the most likely measurement data for machine health condition monitoring. Emerging use of RUL computational techniques by the PHM community has helped in elevating the level of performance accuracy. Transition in the use of conventional finite element method to advanced Bayesian filtering and towards well-proposed PF has been seen in the works of PHM with crack data (An, Choi, & Kim, 2018). Amongst a few notable works using vibration data in the field of PHM, the use of PF and Bayesian data driven approach (Kramti & Ali, 2019) gained much higher appreciation. As complexity in the sensor data increased with the incorporation of non-linearities and non-Gaussianity challenges went higher. Similar works using temperature data was seen for PHM of semiconductor devices (Celaya, Saxena, Saha, & Goebel, 2014) in the industry with extended-Kalman filtering and PF. It would be wise to say that under the circumstance when the measured data was found to be nonlinear and amongst all the data-driven algorithms used in most of the research works, PF finds itself to a much elevated platform when it comes to PHM applications. For a nonlinear noisy dataset the use of advanced signal processing techniques gives rise to an information loss. Thus, necessity arises in handling the noise component in the datasetwhich is another aspect of using PFapproach.

The study on advanced signal processing techniques (Mohanty, 2014) on a degraded data to detect abnormal behaviour by tracking the amount of change in the signal is a first step towards fault detection and estimation. In the use of PF, the authors distinguish two approaches to handle the process non-linearity and noise. Noise can be handled by ignoring the less weighted particles assigned to the updated equation while non-linearity was taken care with the use of resampling (Jouin, Gouriveau, Hissel, Péra, & Zerhouni, 2016).Opportunities with supervised, semi-supervised and unsupervised techniques aimed at anomaly detection in machine learning, and statistical domains have been proposed much before. Research works to identify the abnormal behaviour on several algorithms (Widodo & Caesarendra, 2014) relating to the such proposed techniques has been in wide practice today, such as Support vector machines (SVM), Principle Component Analysis (PCA), Gaussian Mixture Model (GMM), Hidden Markov models (HMMs), Fuzzy Logics besides not restricted to complex structured neural networks (An et al., 2015). However, for a limited class of systems, the application of data-driven maintenance model seems to be more appropriate. Traditional data-driven (An, Choi, & Kim, 2012; Gebraeel, 2006) methods for nonlinear stochastic systems mainly include regression-based models, Wiener processes (Balka, Desmond, & McNicholas, 2009), filtering-based models, risk-based covariance models, and hidden Markov models. Few such case studies with ball bearing, CNC cutting tool and high-speed wind turbine failure showed that vibration signal has a propensity to exhibit strong nonstationary besides nonlinear characteristics and added noise. Therefore, obtaining information regarding faults from these nonlinear and nonstationary signals becomes a must for fault diagnosis. Dynamic Bayesian network (Tobon-Mejia, Medjaher, & Zerhouni, 2012) was used for cutting tool prognosis, and well-known probabilistic framework-based Bayesian methods (Kramti & Ali, 2019) was found suitable for state of the health estimation of wind turbine rotating machinery with corroded ball bearing. Degradation modelling (Saidi, Ben Ali, Bechhoefer, & Benbouzid, 2017) was seen in another such study with high speed wind turbines. Table 11.2 provides a brief idea regarding the type of data required during different modelling approach towards RUL estimation in the field of PHM. This three different class of models will help the user during experimentation to collect the type of data required for computing the RUL.

In utmost real applications, no single CIs stand sensitive to a components failure mode (He, Bechhoefer, Ma, & Zhu, 2012), so a PF-based sequential sampling technique for prognosis is attempted that provides an estimate of the state parameters, thereby assisting in finding the CI a.k.a. RUL for cognitive decision-making to augment the safety. List of practical illustrations in prognostics concerning the use of PF for users in RUL community has been found in the works (Jouin et al., 2016). Amongst the few are mechanical components subjected to crack growth, semiconductor manufacturing, bearing failure and aircraft actuator failure. However, very less results in the study with motors state estimation can be found from the previous works (Jouin et al., 2016) using PF, thereby providing a scope for motor prognosis. Keeping in mind the challenges proposed by the author in applying PF, an attempt towards estimation of model parameters while characterizing the degradation is attempted. The estimated model parameters are then used to find the remaining life of the motor. Figure 11.2 shows the failure state from the present condition/degradation state. The use of different types of algorithms on real-time data during condition monitoring is a challenging task. This increase in the use of algorithms led to an increase in higher unknown parameters, which needs to be wisely chosen for minimal errors. Motor modelling using current as a measurement data can be seen in the works of motor current signature analysis. But the present work uses current

Table 11.2 Data-driven models in predictive maintenance for finding RUL

Similarity model	Degenerate model	Survival model
Uses historical data	Uses known threshold	Uses lifetime data
Compares trend of test data from similar system	Infers past behaviour to predict future	Probability distribution of component failures are used

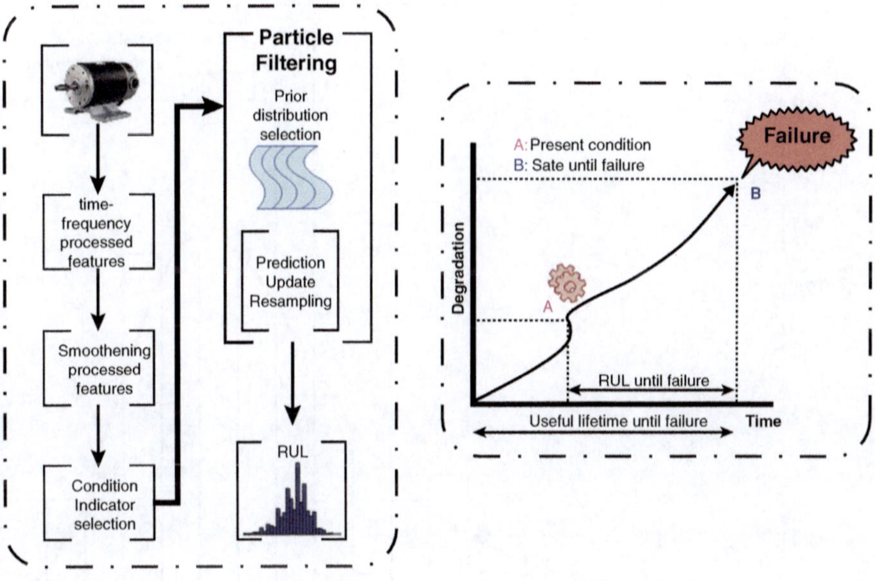

Fig. 11.2 Schematic of RUL prediction

as a measurement data for life prediction of dc motor for stochastic decision-making. Accuracy in the results of the proposed technique is found to be information specific.

The use of estimation theory is in mapping the measured data using an empirically developed model to the estimated state/parameters. For an unobserved variable, '*state*' provides an idea of the system dynamics. Objective function provides the accuracy to the estimator. The true value thus obtained are used to generate the observed data which fulfils the purpose of prediction. Finally, prognosis is carried out with the estimation of the state space model parameters. RUL is then calculated using the estimated model parameters.

11.3 Methodology and Results

For tracking the state of the dynamic system, the mere assumption of a definitive psychological state at every moment is rather superposed by the state space model thereby approximating the belief using a set of particles is what the PF does. In particle filtering, measurement data are collected and conditioned to estimate and update unknown parameters as probability density function via Bayes rule in an attempt to obtain the required posterior state distribution. Thus, is also known to be sequential Monte Carlo method.

$$p(\Theta \mid d) \propto \mathrm{L}(d \mid \Theta)\, p(\Theta) \tag{11.1}$$

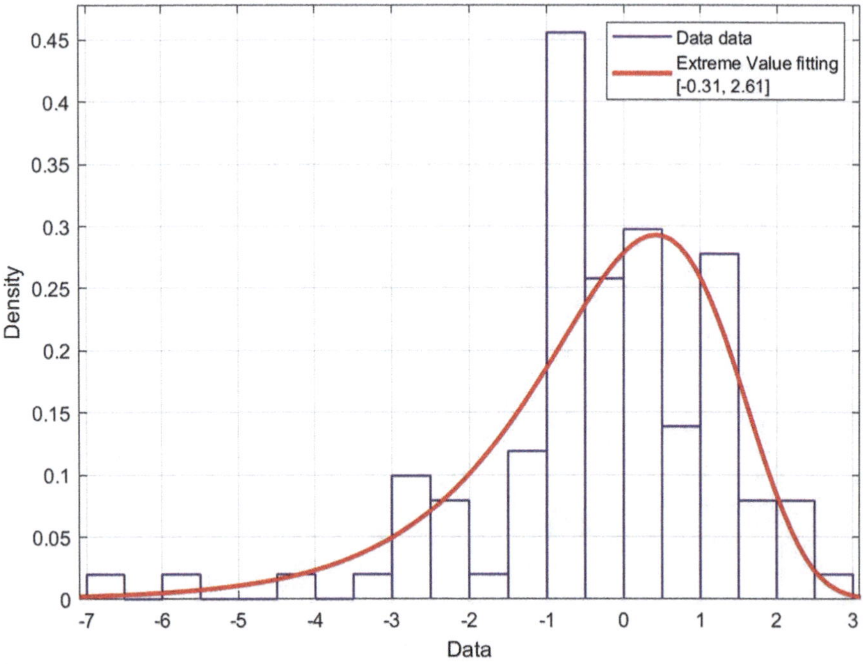

Fig. 11.3 Distribution fitting plot for measured data

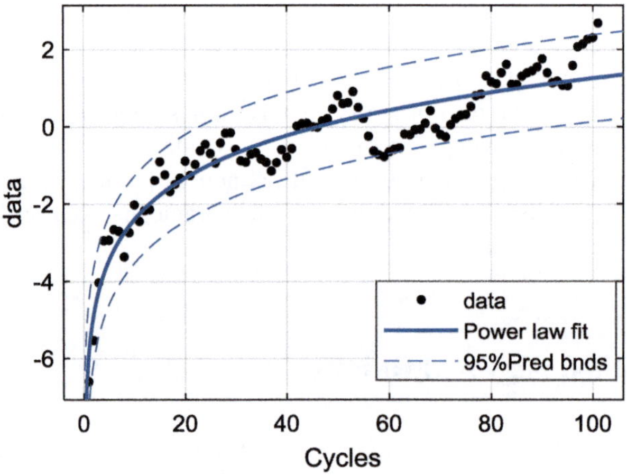

Fig. 11.4 Curve fitting using measured data

Figure 11.3 shows an extreme value distribution fit with measured data and respective scale and shape parameters. The initial estimates are obtained using non-linear regression.

For a degrading motor, '$\Psi(x)$' denotes the measured motor current data as a performance collected during experimentation. An empirical motor degradation model (shown in Fig. 11.4) is expressed using second-order power law model:

$$\Psi(x) = ax^b + c \tag{11.2}$$

PF follows a state transition function 'f' and a measurement function 'm' (An et al., 2012):

$$x_j = f(x_{j-1}, \theta_j, \rho_j)$$
$$d_j = m(x_j, \varepsilon_j)$$

'j' is the time index, 'x_j' is damage state, 'θ_j' is the model parameters and the respective process and measurement noise are denoted using 'ρ_j' and 'ε_j', respectively. Since it follows a recursive relation defined by: $t_j = t_{j-1} + \Delta t$, the model equation with initial degradation state 'd' reduces to the form:

$$x_j = d + ab(d)^{b-1}\Delta x \tag{11.3}$$

The particles (here 5000) having probability information about the unknown parameters $\Theta = \{x, \theta, \sigma\}$ processes in a sequential way using the damage state 'x_k', and model parameters 'a', and 'b'. Figure 11.5 shows the schematic of the PF process for RUL estimation. As time proceeds, the RUL is updated on the newly encountered measurement data. The estimated p'(RUL) for each projected '$j + l$' steps ahead of time 'l' and current time 'j' can be written as:

$$p'(\text{RUL} \le l \mid d_{0:j}) = p'(x_{k+1} \ge \text{thres}|d_{0:j}) \tag{11.4}$$

As soon as the prediction step completes assuming uniform sampling distribution of the prior, it proceeds towards the updating step for the likelihood. For an extreme value distributed, likelihood cumulative distribution is computed, and thereby the weighted samples are normalized again. The process repeats itself a number of times till the data less than the threshold value is encountered. As soon as the state is encountered the process stops and the median value of RUL is triggered (Fig. 11.6). Tabulated (Table 11.3) are the median values of RUL at 10, 50 and 90 percentiles.

11.4 Conclusion

Predictive maintenance is a very promising besides a challenging technique. The advent of Industry 4.0 and the changing standards in the quality and quantity of data from different technological background provided a breakthrough in the use of

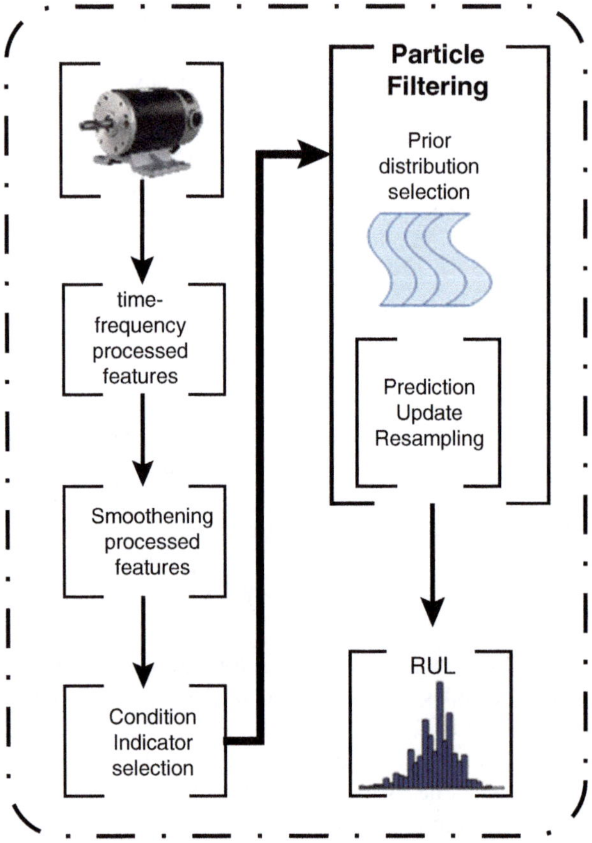

Fig. 11.5 Particle filter-based motor prognostics framework

information processing for estimating the state of health for the system. The use of analytics and machine learning is to preprocess the information from the data, thereby triggering a useful signal to the user needs an unbiased RUL estimation.

A case study depicting the use of sensor data for life estimation of a motor under actual loading conditions is favoured. The use of time-frequency-based signal processing technique is favoured for the particular dataset. Moving mean smoothens the features which favours in finding the correct trend in the data. For a dynamic model, the use of correction step following the prediction step and resampling is an attempt towards reducing the covariance in the results of the study that increases the efficiency in the test results. Results may vary from the typically known ones like shape and scale factor for a respective distribution function of the fitted model, to the most difficult ones like threshold selection or health indicator to serve the purpose of health monitoring.

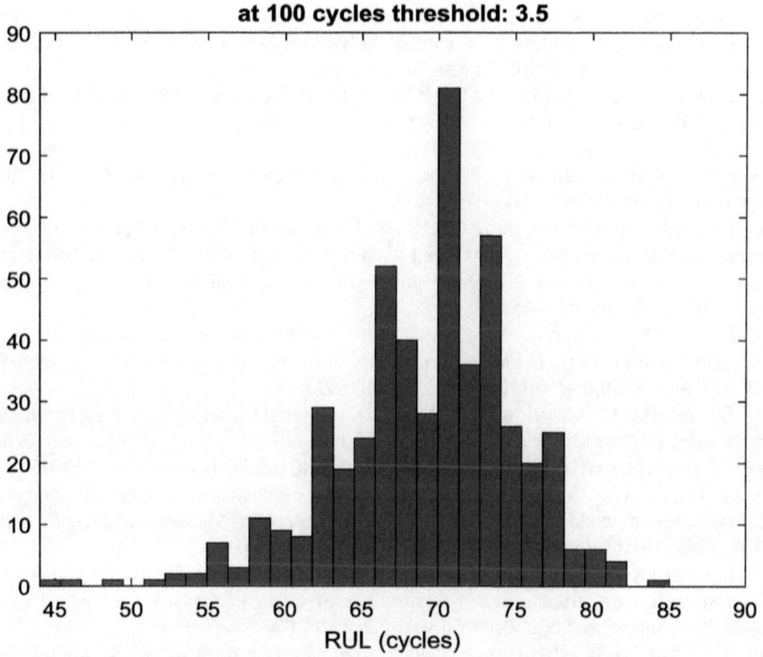

Fig. 11.6 Histogram of RUL plot

Table 11.3 Percentile values of RUL

Percentiles of RUL at 100 cycles		
10prct: 62	Median: 70	90prct: 76

RUL median value obtained from the study thus helps as a robust benchmark indicator for maintenance decision-making. A much appreciated organizational changes, new management approaches with stakeholders for implementing predictive maintenance techniques with advance information technologies will provide a breakthrough to the major obstacle faced in the industry in implementation stage.

References

An, D., Choi, J. H., & Kim, N. H. (2012). A tutorial for model-based prognostics algorithms based on Matlab code. In *Proceedings of the Annual Conference of the Prognostics and Health Management Society 2012, PHM 2012* (pp. 224–232).

An, D., Choi, J. H., & Kim, N. H. (2018). Prediction of remaining useful life under different conditions using accelerated life testing data. *Journal of Mechanical Science and Technology, 32*(6), 2497–2507. https://doi.org/10.1007/s12206-018-0507-z

An, D., Kim, N. H., & Choi, J. H. (2015). Practical options for selecting data-driven or physics-based prognostics algorithms with reviews. *Reliability Engineering and System Safety, 133*, 223–236. https://doi.org/10.1016/j.ress.2014.09.014

Balka, J., Desmond, A. F., & McNicholas, P. D. (2009). Review and implementation of cure models based on first hitting times for Wiener processes. *Lifetime Data Analysis, 15*(2), 147–176. https://doi.org/10.1007/s10985-008-9108-y

Banjevic, D. (2009). Remaining useful life in theory and practice. *Metrika, 69*(2–3), 337–349. https://doi.org/10.1007/s00184-008-0220-5

Celaya, J. R., Saxena, A., Saha, S., & Goebel, K. F. (2014). Prognostics of power mosfets under thermal stress accelerated aging using data-driven and model-based methodologies. In *Proceedings of the Annual Conference of the Prognostics and Health Management Society 2011, PHM 2011* (pp. 443–452).

Gebraeel, N. (2006). Sensory-updated residual life distributions for components with exponential degradation patterns. *IEEE Transactions on Automation Science and Engineering, 3*(4), 382–393. https://doi.org/10.1109/TASE.2006.876609

He, D., Bechhoefer, E., Ma, J., & Zhu, J. (2012). A particle filtering based approach for gear prognostics. In *Diagnostics and prognostics of engineering systems: Methods and techniques* (pp. 257–266). Pennsylvania: IGI. https://doi.org/10.4018/978-1-4666-2095-7.ch013

Jardine, A. K. S., Lin, D., & Banjevic, D. (2006). A review on machinery diagnostics and prognostics implementing condition-based maintenance. *Mechanical Systems and Signal Processing, 20*(7), 1483–1510. https://doi.org/10.1016/j.ymssp.2005.09.012

Jouin, M., Gouriveau, R., Hissel, D., Péra, M. C., & Zerhouni, N. (2016). Particle filter-based prognostics: Review, discussion and perspectives. *Mechanical Systems and Signal Processing, 72–73*, 2–31. https://doi.org/10.1016/j.ymssp.2015.11.008

Kramti, S. E., & Ali, J. B. (2019). *Particle filter based approach for wind turbine high-speed shaft bearing health prognosis*, December 1–6.

Mohanty, A. R. (2014). Machinery condition monitoring: Principles and practices. In *Machinery condition monitoring: Principles and practices*. Boca Raton, FL: CRC Press.

Mohanty, S. N., & Suar, D. (2013a). Decision-making in positive and negative prospects: Influence of certainty and affectivity. *International Journal of Advances in Psychology, 2*(1), 19–28.

Mohanty, S. N., & Suar, D. (2013b). Influence of mood states, group discussion, and interpersonal comparison on change in decision-making and information processing. *International Journal of Decision Sciences, Risk and Management, 5*(2), 101.

Mohanty, S. N., & Suar, D. (2014). Decision making under uncertainty and information processing in positive and negative mood states. *Psychological Reports, 115*(1), 91–105.

Moore, D. W. (2002). Question-order effects additive and subtractive previous research on the measurement of question-order effects. *Public Opinion Quarterly, 66*, 80–91.

Patel, V. L., Kaufman, D. R., & Arocha, J. F. (2002). Emerging paradigms of cognition in medical decision-making. *Journal of Biomedical Informatics, 35*(1), 52–75. https://doi.org/10.1016/S1532-0464(02)00009-6

Pham, H. (1994). *Handbook of reliability engineering*. New York: Springer. https://doi.org/10.1002/9780470172414

Shadlen, M. N., & Kiani, R. (2013). "Decision Making as a Window on Cognition" *Neuron 80*(3): 791–806. https://doi.org/10.1016/j.neuron.2013.10.047

Saidi, L., Ben Ali, J., Bechhoefer, E., & Benbouzid, M. (2017). Wind turbine high-speed shaft bearings health prognosis through a spectral kurtosis-derived indices and SVR. *Applied Acoustics, 120*, 1–8. https://doi.org/10.1016/j.apacoust.2017.01.005

Shmelova, T., Sikirda, Y., Scarponi, C., & Chialastri, A. (2018). Deterministic and stochastic models of decision making in air navigation socio-technical system. *CEUR Workshop Proceedings, 2104*, 649–656.

Si, X. S., Wang, W., Hu, C. H., & Zhou, D. H. (2011). Remaining useful life estimation—A review on the statistical data driven approaches. *European Journal of Operational Research, 213*(1), 1–14. https://doi.org/10.1016/j.ejor.2010.11.018

Tobon-Mejia, D. A., Medjaher, K., & Zerhouni, N. (2012). CNC machine tool's wear diagnostic and prognostic by using dynamic Bayesian networks. *Mechanical Systems and Signal Processing, 28*, 167–182.

Wang, Y., & Ruhe, G. (2007). The cognitive process of decision making. *International Journal of Cognitive Informatics and Natural Intelligence (IJCINI), 1*(2), 73–85. https://doi.org/10.4018/jcini.2007040105

Wang, Z., Busemeyer, J. R., Atmanspacher, H., & Pothos, E. M. (2013). The potential of using quantum theory to build models of cognition. *Topics in Cognitive Science, 5*(4), 672–688. https://doi.org/10.1111/tops.12043

Widodo, A., & Caesarendra, W. (2014). Summary of the recent developed techniques for machine health prognostics. *Rotasi, 16*(1), 21. https://doi.org/10.14710/rotasi.16.1.21-27

Chapter 12
Multiple Task Human Gait Analysis and Identification: Ensemble Learning Approach

Anjali Gupta, and Vijay Bhaskar Semwal

Contents

12.1 Introduction

Walking is one of the most important characteristics of humans. It is a complicated physical activity which is the result of the co-ordination and synchronization of the various body parts together. Human gait refers to the manner in which an individual works. The gait pattern of an individual is influenced by various factors like age, weight, health, gender, social, and geographical conditions. The gait identification finds its application in day to day life in various fields like biometrics, designing of bionic robots, in studying the recovery reports of patients and improving the performance of athletes. A gait cycle comprises of several sub events. Mainly each gait cycle consists of two major phases of the cycle, namely stance and swing phase. Each phase consists of initial, middle, and terminal phase. In stance phase also known as double support phase, both the feet touch the ground. On the other hand, in swing phase also known as single support phase, one limb is on the ground while the other is off the ground to begin the next step. Figure 12.1 shows the stance and swing phase for the normal walk cycle.

W. Hsu et al. in 2018 (Hsu et al., 2018) used multiple wearable sensors for analyzing and classifying the gait of patients with neurological disorders like multiple

A. Gupta · V. B. Semwal (✉)
National Institute of Technology, Bhopal, Madhya Pradesh, India
e-mail: vsemwal@manit.ac.in

S. N. Mohanty (ed.), *Emotion and Information Processing*,
https://doi.org/10.1007/978-3-030-48849-9_12

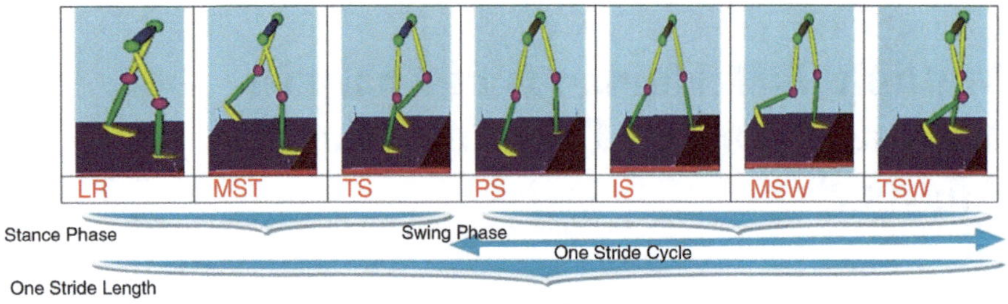

Fig. 12.1 Gait cycle phases

sclerosis, cerebral palsy, and stroke patients. The seven sensors were placed on 20 subjects who produced features to classify different groups. They implemented a number of classification algorithms out of which Multilayer Perceptron (MLP) algorithm outperformed the other classification algorithms (Patil et al., 2019). Clinical gait analysis can also be used for early detection of gait abnormality in brain or neurological disorder subjects (Adil et al., 2016). In 2019, Patil et al. used different types of classification techniques to identify the patients suffering from neurological disorders and stroke patients. The performance of the ELM classifier was found to be the best in terms of accuracy and performance time.

In 2016, Adil et al. proposed surface electromyography (sEMG) signal in order to recognize drop foot of the patient and provide a treatment as they needed for rehabilitation (Adil et al., 2016). They used ELM classifier to classify the sickness and healthy muscles on the leg, and the performance was also compared with the SVM and neural network (NN). The performance of the ELM can be further increased by eliminating its sensitivity to its hyper parameters by presenting a variable length particle swarn optimization algorithm (Ma et al., 2014). This technique was used to optimize the number of hidden neurons corresponding to input weights and biases of elm. Another approach was proposed by Guo, Wu, Shen, Zhang, and Zhang (2019) to identify gait disorders in Parkinson's disease by using machine learning algorithms in 2019. This model was used to classify four different types of gait abnormality. Another novel approach was proposed by Mekruksavanich and Jitpattanakul, in 2019, where wearable sensors on smartphones were used to collect data. The study was made to classify gait patterns for three different activities like walking upstairs, walking downstairs, and walking on the floor (Mekruksavanich & Jitpattanakul, 2019) using different types of learning techniques.

Gait can also be used as a biometric characteristic which can find its application in various fields like surveillance and forensics (Anusha & Jaidhar, 2019; Gowtham Bhargavas, Harshavardhan, Mohan, Nikhil Sharma, & Prathap, 2017; Xu, Makihara, Yagi, & Lu, 2019). A number of techniques have been proposed like utilizing mutual information obtained from a query and gallery sample. This method uses the ROI extracted from gait energy image to perform classification and identification. In another approach, Bhargavas et al. proposed a method in which subject is identified from video frame by extracting skeleton information obtained from sensors. Another application of gait analysis is in clinical and medical purposes where the patients suffering from neurological disorders can analyze their gait abnormalities and can design personalized treatment (Papavasileiou et al., 2017). Human gait classification is also an important area of research interest where a number of techniques are implemented to study and analyze the gait pattern for different human locomotion activities like walking, running, and climbing. Chen et al. (Bovi, Rabuffetti, Mazzoleni, & Ferrarin, 2011) in 2018 proposed deep convolutional neural networks (DCNNs) based on multistatic radar micro-Doppler signatures for gait classification (Mohanty & Suar 2014; Nandi et al., 2016). A new gait-based gender classification method based on kinect sensor was proposed by Ahmed et al. (2017) (Semwal et al., 2016a). The other techniques used previously for gait classification approach are using different covariate factors, duty factor, and dictionary learning-based approach (Ahmed & Sabir, 2017; Chen, Li, Fioranelli, & Griffiths, 2018; Poschadel, Moghaddamnia, Alcaraz, Steinbach, & Peissig, 2017).

This chapter is organized as follows. Section 12.2 describes dataset and methodology used. Section 12.3 briefs about the different classifiers used in the proposed work. Section 12.4 analyzes and compares the results achieved by different machine learning algorithms. Section 12.5 deals with the conclusion and future enhancements.

12.2 Proposed Work

12.2.1 Dataset Description and Methodology

The proposed work has used a multiple-task gait analysis approach: kinematic, kinetic, and EMG reference data for healthy young and adult subject dataset. The dataset contains data from the various tasks: walking at natural speed (N), walking very slow (XS), walking slow (S), walking medium (M), walking fast (L), walking on toes (T), walking on heels (H), stair ascending (U), and stair descending (D). Table 12.1 consist of 909 rows and 7 columns, with the last column labeled as OUTPUT shows the various task and the rest of them representing the values of different features used to classify the data.

Above we can see Table 12.1 which gives details of the database shape and the class distribution, along with a statistical analysis about the database.

The proposed system consists of sequence of steps starting from dataset collection to training and classification of the dataset. The dataset consists of number of features containing irrelevant information. So, in order to refine and pre-process the dataset, Principal Component Analysis (PCA) algorithm is applied to extract the significant features. This will also help in improving the performance of the proposed system as it will reduce the training and classification time.

After extracting the best features and ignoring the irrelevant details, the final dataset is obtained. The dataset is then divided into training and testing group on which different classification algorithms are applied. The different classification algorithms like KNN, SVM, and ELM were applied to classify the data. The classification results obtained from the training and testing pair are then evaluated and compared to compute the accuracy and efficiency of each of the classifier used (Fig. 12.2).

Table 12.1 Dataset description

	Pelvic Ant/Posterior	Hip Flex/Extension	Hip Ad/Abduction	Hip Int/External rotation	Knee Flex/Extension	Ankle Dorsi/Plantarflexion	OUTPUT
0	−2.8	27.9	0.5	0.3	8.8	−21.9	0
1	−2.8	27.8	0.9	0.6	10.2	−22.5	0
2	−2.8	27.7	1.4	1.0	11.5	−23.1	0
3	−2.7	27.6	1.9	1.4	12.9	−23.7	0
4	−2.7	27.5	2.5	1.8	14.5	−24.1	0

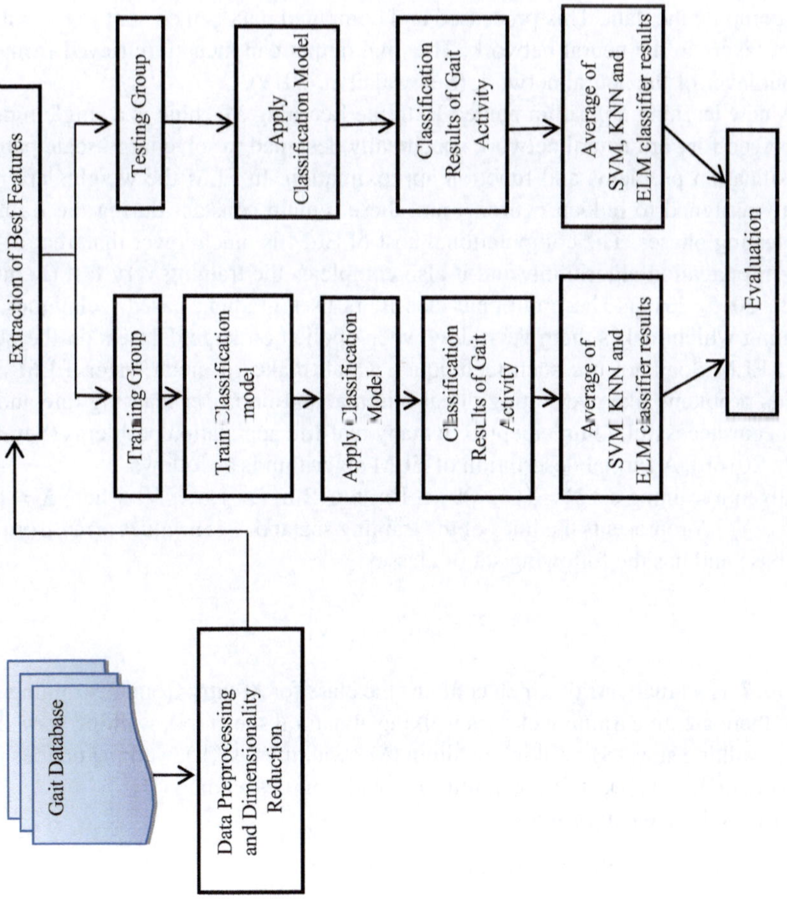

Fig. 12.2 Proposed architecture

12.3 Algorithm

12.3.1 Extreme Learning Machine Algorithm

Artificial neural networks in particular Feed forward Neural Network are multi-layer fully connected neural networks which consist of an input layer, multiple hidden layers, and an output layer (Fihland & Moeslund, 2007; Ng, Tong, Tan, Yap, & Abdullah, 2010). Each layer is composed of several processing units which process and compute the data. This processed and computed data is then sent to the subsequent layers in the neural network. The final output can then be retrieved from the output layer of the neural network (Semwal et al. 2019).

A new learning algorithm named Extreme Learning Machine is a single hidden layer feed forward neural network specifically designed to solve large-scale pattern classification problems and function approximation. In ELM the weights are randomly assigned to hidden neurons, and these remain constant during the training and testing phases. The computational cost of ELM is much lower than that of the back propagation algorithms, and it also completes the training very fast (Semwal et al., 2015, 2017). The traditional classifiers uses gradient based techniques for learning which makes them very slow when applied on large datasets on the other hand ELM does not use such techniques which makes it much faster. ELM also avoids problems like over fitting, local minima, and improper learning rate and so has been successfully implemented in many real-life application problems (Semwal et al., 2016b). A formal description of ELM algorithm is as follows:

Given a training set $\aleph = \{(x_i, t_i) | x_i \in \text{Rn}, t_i \in \text{Rm}, i = 1, \ldots, N\}$ where $X = \{X_1, X_2, \ldots, X_N\}$ N represents the lines of the training set, and X_i is a values array from the database and has the following set of classes:

$$T = \{T_1, T_2, \ldots, T_N\}$$

where, T_i is a binary array which contains the class for X_i entry from the training set. Here there are nine training classes walking at natural speed (N), walking very slow (XS), walking slow (S), walking medium (M), walking fast (L), walking on toes (T), walking on heels (H), stair ascending (U), and stair descending (D).

ELM implementation steps:

1. Generate the weights matrix for the input layer

$$W = \begin{bmatrix} rand & \cdots & rand \\ \vdots & \ddots & \vdots \\ rand & \cdots & rand \end{bmatrix}, \tag{12.1}$$

2. Calculate the hidden layer output matrix

$$H = W * X \tag{12.2}$$

3. Calculate the Moore–Penrose pseudoinverse

$$G^{+} = \left(G^{\tau} * G\right)^{-1} * G^{\tau}, \tag{12.3}$$

4. Calculate the output weight matrix beta

$$\beta = H^{+} * T. \tag{12.4}$$

Essential considerations of ELM

- High accuracy
- Least user intervention
- Real-time learning (in seconds, milliseconds, even in microseconds)

12.3.2 Support Vector Machines (SVM)

Support Vector Machine (SVM) is another simple machine learning algorithm which can be used for both regression and classification problems. Generally, it is used for classification task. The main task of the SVM is to find optimal separating hyper plane in an n-dimensional plane to classify data points. Many hyper planes are possible when one needs to classify the data, but the aim is to select that optimal hyper plane which has maximum distance between data points of the two classes. This optimal separating hyper plane obtained is then further used to classify future data points with more accuracy and precision. SVM works really well in high dimensional spaces and is also memory efficient. But the performance of this algorithm usually degrades when one has noisy and large datasets as the computational time would increase considerably.

12.3.3 k-Nearest Neighbor (KNN)

The k-Nearest Neighbor Algorithm is a supervised machine learning algorithm which can be used in classification as well as regression problems. This algorithm classifies objects on the basis of the distance of the query object to the training samples in the feature space. The selection of the parameter k strongly influences the performance of this algorithm therefore proper selection of this parameter is of utmost importance. The KNN algorithm takes less training time and performs well for the large datasets. However, its computational costs are high as it takes into consideration and computes the distance of each query to all the training samples. This algorithm finds its applications in various fields like text mining, agriculture, finances like stock market analysis, and credit rating.

12.3.4 Principal Component Analysis (PCA)

Principal Component Analysis refers to reducing the dimension in feature space. Dimensionality reduction can be achieved either by feature extraction or feature elimination. In feature elimination, few features are eliminated which would also eliminate information from these eliminated features which could play significant role in producing better results. In feature extraction new independent features are created in specific way from the existing ones, so that insignificant variables can be ignored without ignoring the useful information from them. So PCA examines and identifies the dataset and computes similarities and differences within the data. The covariance matrix generated by PCA gives measure of how one variable is associated with the other variable. The Principal Component Analysis therefore reduces the correlated features which enhances algorithm performance, reduces over fitting, and improves visualization.

12.3.5 Ensemble Learning

Ensemble learning technique used to avoid over fitting by reducing the complexity of model. It used to consider the average performance of different classifier and different hyper parameter. So, if any classifier performs poor and another performs better, the ensemble learning used to combined all results. The average of all classifier will provide the much needed generality of model by reducing variance and model complexity.

12.4 Results and Discussions

The experiments were performed on Windows 10 machine with 3.20 GHz i8 8700U CPU and 8.0 GB RAM. The dataset consists of the following features like Pelvic Ant/Posterior Tilt, Hip Flex/Extension, Hip Ad/Abduction, Hip Internal/External Rotation, Knee Flexion/Extension, and Ankle Dorsi/Plantar flexion. The Principal Component Analysis algorithm was used to reduce the dimension of the dataset in order to reduce the computational time of classification. The explained variances calculated among these features are shown in Table 12.2.

ELM, SVM, KNN, and PCA were implemented using scikit-learn python module. The accuracy of all the classification techniques is computed to compare the performance of each of the algorithms implemented. The results also incorporated precision, recall, F1 score, and support for all the labeled outputs. Confusion matrix is also displayed for all the above-mentioned classification techniques. Accuracy is a performance measure which can be defined as the ratio of correctly predicted observation to the total observation. Precision is the ratio of the predicted positive

Table 12.2 Variance computed among different features

S. No	Feature name	Variance
1	Pelvic Ant/Posterior Tilt	0.30811954
2	Hip Flex/Extension	0.25931277
3	Hip Ad/Abduction	0.19041871
4	Hip Internal/External Rotation	0.10793897
5	Knee Flexion/Extension	0.08192927
6	Ankle Dorsi/Plantar flexion	0.05228074

Table 12.3 SVM classification report

	Precision	Recall	F1score	Support
0	0.67	0.52	0.58	27
1	1.00	1.00	1.00	17
2	0.59	0.81	0.68	21
3	0.94	0.67	0.78	24
4	0.94	1.00	0.97	17
5	1.00	1.00	1.00	18
6	1.00	1.00	1.00	18
7	1.00	1.00	1.00	19
8	0.72	0.86	0.78	21

observation to the total predicted positive observation. Recall is the ratio of the true positive observation to the total actual positive observation. F1 score on the other hand is the weighted average of the precision and recall. The dataset consists of nine classes, namely walking at natural speed (0), walking very slow (1), walking slow (2), walking medium (3), walking fast (4), walking on toes (5), walking on heels (6), stair ascending (7), and stair descending (8). There are a total of 909 samples across all nine classes with their individual numbers given under "OUTPUT" column in the classification report table (Table 12.3). Figure 12.3 shows the graphical representation of the implementation of PCA algorithm on the dataset.

In KNN classifier, the selection of the parameter k strongly influences the performance of this algorithm; therefore, proper selection of this parameter is of utmost importance. Figure 12.4 shows the accuracy of this classifier on the various values of k. The results show that the KNN algorithm provides the highest accuracy when $k = 3$.

Table 12.4 represents the classification report for KNN. The overall accuracy of the KNN classifier was found to be 79%.

The overall accuracy of the ELM classifier was found to be 84.06%. To improve the overall classification accuracy of different activity gait, the performance of all different classifier has been combined for different hyper parameter. The average of all three classifier has given overall 82% accuracy in four different trials (Fig. 12.5). Table 12.5 shows the overall classification accuracy report of different classifier.

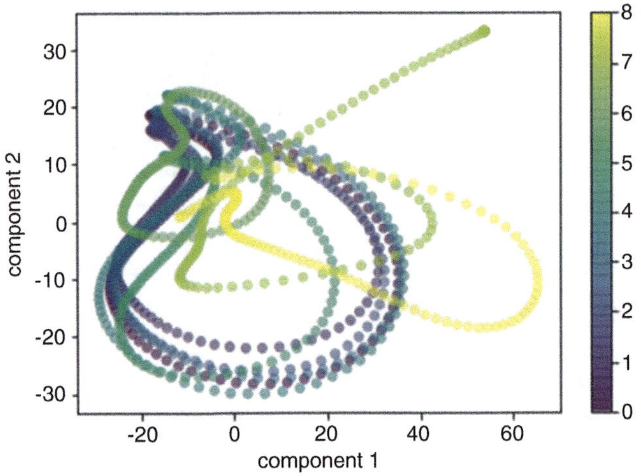

Fig. 12.3 PCA applied to the dataset

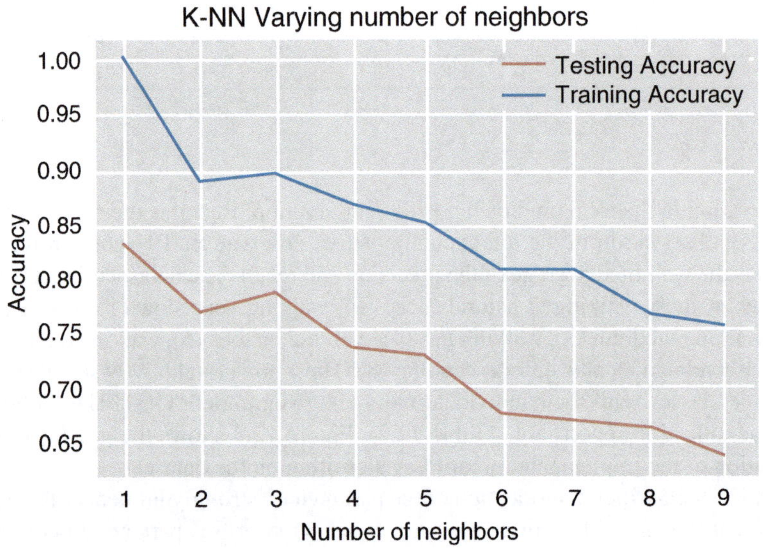

Fig. 12.4 KNN classifier

12.5 Conclusions

This chapter compared the performance of ELM with the other classifiers namely SVM and KNN for multiple task human gait analysis and identification. The accuracy of proposed system is 84.06%. The results indicate that ELM gives better classification results within less time. This chapter proposed a gait recognition method for different walking patterns using ELM classification, and the results were also

Table 12.4 KNN classification report

	Precision	Recall	F1 score	Support
0	0.32	0.57	0.41	40
1	0.93	0.95	0.94	40
2	0.73	0.54	0.62	41
3	0.47	0.34	0.39	41
4	1.00	0.76	0.86	41
5	1.00	0.95	0.97	40
6	1.00	1.00	1.00	41
7	0.95	1.00	0.98	40
8	1.00	0.97	0.99	40

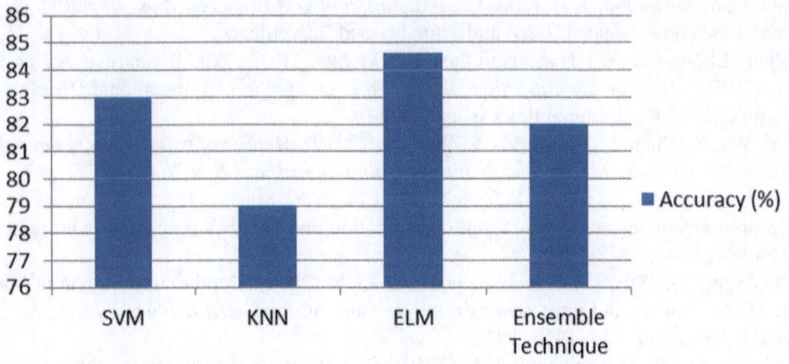

Fig. 12.5 Accuracy classification report

Table 12.5 Overall classification accuracy report of different classifier

S. No	Classifier	Accuracy (%)
1	SVM	83
2	KNN	79
3	ELM	84.06
4	Ensemble technique	82

compared with the other classification techniques like SVM and KNN. In this work, our algorithm identified nine different types of walking pattern behavior from the given dataset in less time. In future, autoencoders can also be replaced by PCA for improved results. The results therefore proved that the ELM classifiers are fast machine classifiers which find its place in real world applications. The paper has also implemented the ensemble learning by providing the combined average results of different classifier. It provides the generic solution and remove the dependency on more on hyper parameter and data dependence. The ensemble technique has reduced the variance and model complexity.

References

Adil, S., et al. (2016). Extreme learning machine based sEMG for drop-foot after stroke detection. International Conference on Information Science and Technology.

Ahmed, M. H., & Sabir, A. T. (2017). *Human gender classification based on gait features using kinect sensor*. IEEE International Conference on Cybernetics (CYBCONF).

Anusha, R., & Jaidhar, C. D. (2019). *An approach to speed invariant gait analysis for human recognition using mutual information*. TENCON, IEEE Region 10 International Conference.

Bovi, G., Rabuffetti, M., Mazzoleni, P., & Ferrarin, M. (2011). Dataset: A multiple-task gait analysis approach: Kinematic, kinetic and EMG reference data for healthy young and adult subjects. *Gait Posture, 33*(1), 6–13.

Chen, Z., Li, G., Fioranelli, F., & Griffiths, H. (2018). Personnel recognition and gait classification based on multistatic micro-Doppler signatures using deep convolutional neural networks. *IEEE Geoscience and Remote Sensing Letters, 15*(5), 669–673.

Fihland, P., & Moeslund, T. B. (2007) *Classification of gait types based on the duty-factor*. IEEE Conference on Advanced Video and Signal Based Surveillance.

Gowtham Bhargavas, W., Harshavardhan, K., Mohan, G. C., Nikhil Sharma, A., & Prathap, C. (2017). *Human identification using gait recognition*. International Conference on Communication and Signal Processing (ICCSP).

Guo, Y., Wu, X., Shen, L., Zhang, Z., & Zhang, Y. (2019). *Method of gait disorders in Parkinson's disease classification based on machine learning algorithms*. New York: IEEE.

Hsu, W. C., Sugiarto, T., Lin, Y. J., Yang, F. C., Lin, Z. Y., Sun, C. T., ... & Chou, K. N. (2018). Multiple-wearable-sensor-based gait classification and analysis in patients with neurological disorders. *Sensors, 18*(10), 3397.

Ma, X., Wang, H., Xue, B., Zhou, M., Ji, B., & Li, Y. (2014). Depth-based human fall detection via shape features and improved extreme learning machine. *IEEE Journal of Biomedical and Health Informatics, 18*, 1915–1922.

Mekruksavanich, S., & Jitpattanakul, A. (2019). *Classification of gait pattern with wearable sensing data*. New York: IEEE.

Mohanty, S. N., & Suar, D. (2014). Influence of mood states on decision making under uncertainty and information processing, *Psychological Reports, 115*(1), 44–64.

Nandi, G. C., et al. (2016). Modeling bipedal locomotion trajectories using hybrid automata. In *IEEE Region 10 Conference (TENCON)*. New York: IEEE.

Ng, H., Tong, H.-L., Tan, W.-H., Yap, T.-V., & Abdullah, J. (2010). *Gait classification with different covariate factors*. International Conference on Computer Applications and Industrial Electronics.

Papavasileiou, I., Zhang, W., Wang, X., Bi, J., Zhang, L., & Han, S. (2017). *Classification of neurological gait disorders using multi-task feature learning*. IEEE/ACM International Conference on Connected Health: Applications, Systems and Engineering Technologies (CHASE).

Patil, P., et al. (2019). *Clinical human gait classification: Extreme learning machine approach*. First International Conference on Advances in Science, Engineering and Robotics Technology 2019 (ICASERT 2019).

Poschadel, N., Moghaddamnia, S., Alcaraz, J. C., Steinbach, M., & Peissig, J. (2017). *A dictionary learning based approach for gait classification*. 22nd International Conference on Digital Signal Processing (DSP).

Semwal, V. B., et al. (2015). Biometric gait identification based on a multilayer perceptron. *Robotics and Autonomous Systems, 65*, 65–75.

Semwal, V. B., et al. (2016a). Design of vector field for different subphases of gait and regeneration of gait pattern. *IEEE Transactions on Automation Science and Engineering, 15*(1), 104–110.

Semwal, V. B., et al. (2017). Robust and accurate feature selection for humanoid push recovery and classification: Deep learning approach. *Neural Computing and Applications, 28*(3), 565–574.

Semwal, V. B., et al. (2019). Human gait state prediction using cellular automata and classification using ELM. In *Machine intelligence and signal analysis* (pp. 135–145). Singapore: Springer.
Semwal, V. B., et al. (2016b). An optimized feature selection technique based on incremental feature analysis for bio-metric gait data classification. *Multimedia Tools and Applications, 76*(22), 24457–24475.
Xu, C., Makihara, Y., Yagi, Y., & Lu, J. (2019). Gait-based age progression/regression: a baseline and performance evaluation by age group classification and cross-age gait identification. *Machine Vision and Applications, 30*, 629–644.

Chapter 13
Employing Machine Learning for Multi-perspective Emotional Health Analysis

Monika Mangla, Rakhi Akhare, Sanjivani Deokar, and Vaishali Mehta

Contents

13.1 Introduction

In this era of digital revolution, people are surrounded by smart devices and gadgets. Revolution in the field of IoT (Internet of Things) and IoE (Internet of Everything) has further resulted in gaining control of human life. Although these devices were earlier designed to be controlled by humans, but in recent few years, it has been observed that humans are getting driven by these devices. However, this excessive usage of technology in human life has resulted in several concerning issues related to health. Generally, the human health is perceived to be the state of the physical health ignoring the state of mental health. This ignorant attitude toward mental health is supported by the lack of research in this field. However, during the past few years, researchers are focusing on studying and understanding the emotional health of patients. There is no universal agreement on the definition of emotion. However, emotion may be defined in relation to a list of predefined descriptors such as anger, happiness, and sadness.

Here, it is worth mentioning that the emotional health of ordinary people is also equally concerning and thus must be focused upon (Tivatansakul & Ohkura, 2014). Now, it must be realized that society is in dire need of developing systems that aid

M. Mangla (✉) · R. Akhare · S. Deokar
Computer Science and Engineering Department, Lokmanya Tilak College of Engineering, Navi Mumbai, India

V. Mehta
Karnal Institute of Technology and Management, Karnal, India

S. N. Mohanty (ed.), *Emotion and Information Processing*,
https://doi.org/10.1007/978-3-030-48849-9_13

people to get rid of mental stress and negative emotions (Ambarkar & Akhare, 2020). Various researchers have devised different strategies to improve emotional health of patients by providing effective, intelligent, and attractive healthcare systems.

All these proposed systems understand that the most crucial part of emotional health management is analysis and recognition of real-time emotions which is capable of understanding user emotions in minimum time, so that appropriate corrective therapies could be activated or suggested. Recognition of emotional health is a challenging part as it involves recognition of facial expressions, speech, text, or biological signals for enhanced accuracy (Tivatansakul & Ohkura, 2014). According to a popular psychological theory, emotion may be defined as a complex psychological state that involves three distinct components: a subjective experience, a physiological response, and a behavioral or expressive response. Hence, it is evident that analyzing emotional health is extremely challenging. Despite the involved complexities, researchers have been attempting to devise efficient approaches owing to its widespread application in several areas. Some of these application examples are social media analysis, measuring satisfaction level of customers, chatbots integration, sophisticated robot communication, call center performance monitoring, etc. Emotional health analysis is also referred as affective computing as it measures the affect (emotions) (Hakak, Mohd, Kirmani, & Mohd, 2017). The authors of this chapter aim to present the employment of machine learning (ML) in analysis of emotional health. The chapter has been organized as follows:

The introduction has been presented in Sect. 13.1. The basics of emotional health analysis have been presented in Sect. 13.2. Section 13.2 also discusses all principle perspectives of emotional health analysis. Comparative study of these different perspectives have been presented in Sect. 13.3. Section 13.4 presents some promising case studies. Section 13.5 is dedicated to conclusion and future directions of research in the domain.

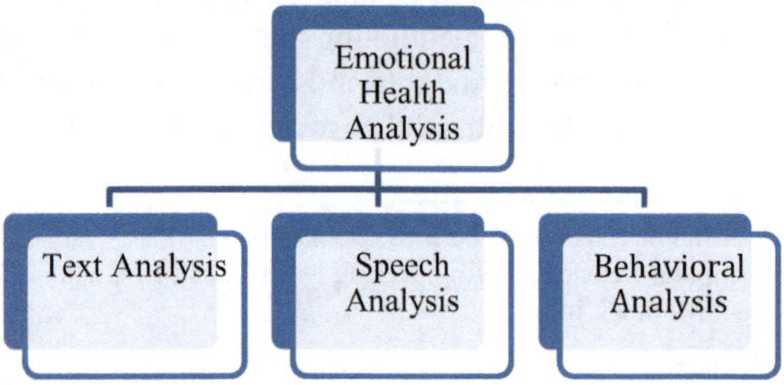

Fig. 13.1 Principle components of Emotional Health Analysis

13.2 Emotional Health Analysis

As stated earlier, emotional health analysis comprises analyzing speech, text, and behavior of the concerned which has been illustrated in Fig. 13.1. Various researchers have applied distinct approaches to analyze different components of emotional health. Among several approaches, Artificial Intelligence (AI) has established itself as a promising and efficient tool (Ramalingam, Pandian, Jaiswal, & Bhatia, 2018).

AI employs various techniques to analyze facial expressions, body movements, speech, and textual information in order to analyze emotional health of a human. For all these components of emotional health, there exists a challenge to identify which data should be used for feature extraction and how to overcome the evolution or change in these components. Some additional challenges involve selection of emotion indicators for speech, extraction of contextual data, labeling of emotions, and selection of machine learning classifier. There have been several computational approaches to implement an emotional classifier. These classifiers have been broadly classified into lexicon based approach and the machine learning approach as shown in Fig. 13.2.

Emotion lexicon contains textual units annotated with emotional labels and work on lexicons. ML approaches use different ML algorithms to train the system and map a function for emotion classification. Some promising approaches of machine learning are supervised learning, unsupervised learning, and semi-supervised learning as demonstrated in Fig. 13.2. The various components of emotional health analysis have been discussed in subsequent subsections.

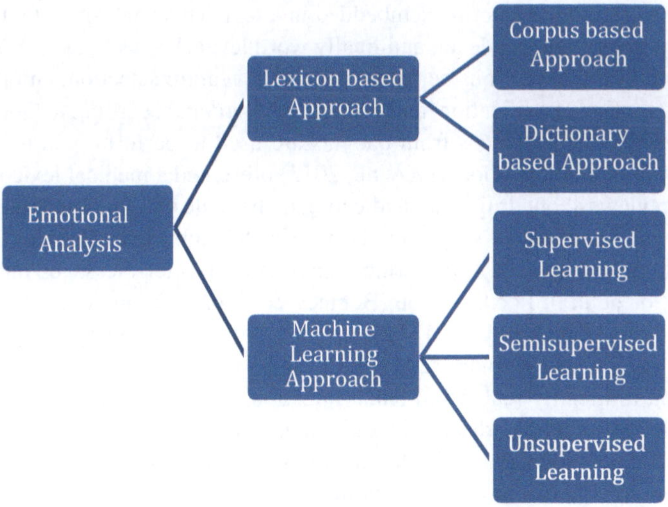

Fig. 13.2 Hierarchy of computational approaches for emotional analysis

13.2.1 Emotion Analysis Using Text Processing

Language is a vital method of communication to express various emotions like love, anger, anxiety etc. This subsection considers the human emotion expressed through text. Humans can easily interpret the textual message, but it remains a challenging task for computers to interpret and understand the text (Khan & Ejaz, 2016). To address this challenge, human–computer interaction (HCI) provides a big relief. Humans provide massive data on the Internet every second which is challenging to digitize. Various emotions like joy, happiness, anger, surprise, fear, and love are expressed by different kinds of words. Some words also belong to the territory category of emotions, thus obscuring the recognition (Mohanty & Suar, 2014). This categorization is used to perform sentiment analysis. According to Carrillo-de-Albornoz, Vidal, and Plaza (2018), subjects are classified into positive, negative, and neutral personalities (Mohanty & Suar, 2013). Broadly, the emotional analysis through text processing has been classified into two classes, viz. hard sensing and soft sensing (Khan & Ejaz, 2016). In hard sensing, brain signals, heart rate, etc. are analyzed. On the other hand, soft sensing detects emotions through emails, text messages, and social website interactions.

Text processing is a significant component of analyzing emotional health. Apart from monitoring a patient's emotional health, it has also been applied to analyze satisfaction levels of customers and employees in big companies. For its implementation in analyzing emotional health of patients, the daily observations of a patient are documented in terms of degree of pain or any other discomfort experienced (Deng, Stoehr, & Denecke, 2014). This documented information is further processed in order to analyze the patient's health that eventually supports and guides treatment process.

Natural language processing (NLP) and few other computational techniques are used to determine the emotions embedded in a text. This analysis of text is done at document level, sentence level, and finally word level (Hakak et al., 2017). At each level of analysis, the actions performed are NLP, feature selection, emotion identification, emotion classification, and evaluation in order. According to Goeuriot et al. (2012), subjective sentences from patients are used to perform polarity classification. Similarly, authors (Goeuriot et al., 2012) presented a medical lexicon containing user reviews about drugs and medication in the scale of 0 to 10 and thus performs polarity classification on subjective issues. In the same line, Bobicev, Sokolova, Jafer, and Schramm (2012) performed analysis of Twitter messages into positive, negative, or neutral. For the same, Bobicev et al. (2012) employed Naive Bayes, Decision trees, KNN, and SVM algorithms on different word bags.

Lim, Tucker, and Kumara (2017) proposed an unsupervised ML model that is competent to identify real-world latent infectious diseases by mining social media data. A latent infectious disease is a communicable disease that has not yet been formalized by national public health institutes like COVID-19 (Lim et al., 2017). In such latent infectious diseases, symptoms are unknown. Lim et al. (2017) have presented a study to discover latent infectious disease through social media messages where users share their experience about symptoms, body pain locations etc. For discovering symptoms of latent disease, the suggested approach uses unsupervised sentiment analysis and social media content pertaining to the same disease. For instance, a social post like "I had a headache the past 2 days, feeling better now because drugs" expresses positive feelings as the handler has gotten rid of the headache. However, if a message does not express body part of feeling, it cannot be classified as an indicator of symptoms. An example of such message is "I hate experiencing mean friendship."

Ji, Chun, Wei, and Geller (2015) also study Twitter about diseases' outbreak. It has been realized that it is important to understand and analyze the public sentiments during outbreaks of Ebola in Africa and measles on the West Coast of the USA. The same can also be realized during the recent pandemic outbreak of coronavirus in the whole world. Understanding the unparalleled demand for monitoring the social data is supported by developing a gauge that evaluates the measures of concern (MOC) using Twitter (Ji et al., 2015). Here, the tweets are classified into personal tweets and news tweets. The personal tweets are classified into negative or non-negative tweets. The classification of tweets into personal and non-personal tweets is a novel concept in Ji et al. (2015) as traditional Twitter sentiment analysis methods do not distinguish among personal and non-personal tweets. This classification model is demonstrated in Fig. 13.3.

Ji et al. (2015) have developed an intelligent system Epidemic Sentiment Monitoring System (ESMOS) which collects Twitter data related to public health. Thereafter, ESMOS classifies the data into different sentiment categories and calculates the degree of concern. ESMOS can also represent the Intensity map using various visual tools which enables understanding the spatial distribution and concentration of public concerns. The data is classified into labeled classes using keyword spotting technique in (Shivhare & Saritha, 2014) where words like sad, angry, fearful, and surprised are used for classification. Another promising approach for this classification is machine learning where previously trained classifiers are used to identify which category the input text belongs to. The same can also be achieved using integration of techniques in hybrid methods which achieves enhanced accuracy. These approaches for tweets classification bear some limitations also. Some of these limitations are ambiguity in keywords, complex nature of emotions, incapability of recognizing sentences in the absence of keywords, lack of linguistic information, determination of emotion indicators, etc. Ramalingam et al. (2018)

Fig. 13.3 Illustration of tweet classification model

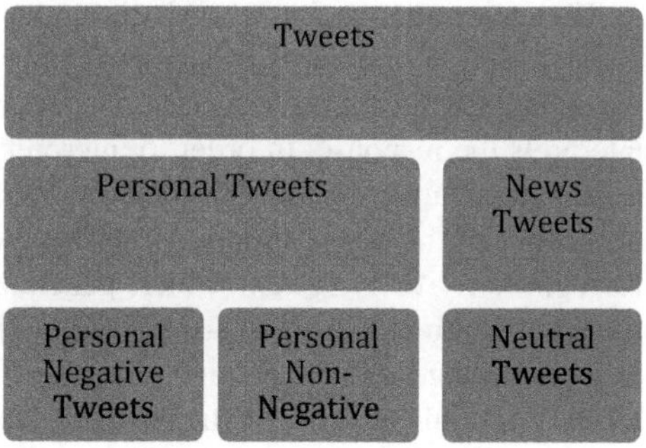

have attempted to address these limitations by presenting a method that uses algorithms of Support Vector Machine and thus providing a better and accurate solution.

Bosubabu Sambana (2017) have proposed a method to assess the behavior of aspiring candidates for the recruitment process. It works by analyzing the candidate's tweets in order to assess his emotions and polarity as entire tweets provide a large dataset, and thus extraction of personality traits becomes more effective and meaningful. Such a model is competent to predict the candidate's behavior in the condition of work pressure and competition and thus helps the recruiting companies to hire the right person.

13.2.2 Emotion Analysis Using Speech Processing

Speech is an unparalleled mode of communication to express emotions among humans. Each speaker has its own voice quality which conveys vital information like emotion and attitude of the speaker. Also, this mode of communication is important as it is able to analyze emotions of the society comprising a large portion of uneducated people who are unable to express their emotions through text. Another motivation for research in the domain of emotion analysis through speech is its instantaneous and real nature. As a result, rigorous research is taking place in this direction for proper psychometric analysis of people in order to understand their psychological problems like bipolar disorders, severe depression, anxiety and schizophrenia etc. It also carries a vital significance in analyzing the emotional health of patients which aids the treatment process (Vij & Pruthi, 2018) (Mohanty, Pratihar, & Suar, 2015). Vij and Pruthi (2018) have developed a psychometric analyzer that analyzes the emotional health of patients based on medical history and recorded communication. Here, the emotional health is analyzed in terms of various components viz. intensity, emotions, polarity and subjectivity. Kerkeni et al. (2019) have defined the emotional health analysis through speech as a four step process. These four steps are as follows:

Collection of voice samples
Feature extraction
Feature selection
eClassification

This emotional health analysis through speech has also been accepted in automatic detection of dementia diseases (Mirheidari, Blackburn, & Walker, 2019). For the same, the health professional asks specific questions to the patient and then interprets the response. In order to interpret the speech, the audio file is processed using some diarization tool that identifies the speech portion and the speaker. Further the output of the diarization tool and the audio file is passed to Automatic Speech Recognition (ASR) system which generates a string of words spoken by each speaker (health professional and patient) (Mirheidari et al., 2019). These are further analyzed using natural language processing and other similar approaches to evaluate the emotional health of patients. Chen, Yang, Hao, Mao, and Hwang (2017) have also developed a model for emotional health analysis leveraging the deep learning and ML algorithms.

Additionally, Davletcharova, Sugathan, Abraham, and James (2015) have interestingly established that emotions influence the heart rate and nervous system also. This principle is employed to assess the emotional state of a person through heart rate as speech is also influenced by transitions in heart rate. For instance, in case of a negative stimuli, the heart rate decelerates rapidly in comparison to positive stimuli. The task of emotional analysis through speech recognition has also been researched in (Nwe, Foo, & De Silva, 2003) where the authors used a discrete Hidden Markov Model (HMM) for the classification of emotions into six labeled classes. Here, authors have used a database of 60 emotional utterances from 12 speakers for training the model. This line of research has been carried forward by few other researchers by employing other promising approaches. The readers can refer to Davletcharova et al. (2015) for detailed analysis of the same.

13.2.3 Emotion Analysis Using Behavior Perception

Emotional health of a person can also be analyzed from his or her behavior in addition to text and speech. Unlike text and speech, the behavior of a person is the most authentic and natural perspective of expression of emotions and intentions. Hence, the behavior of a person should be closely monitored and analyzed in order to understand emotional state of a person. The behavior of a person is mainly demonstrated by his facial expression that mainly studies contraction and expression of facial muscles and eyes. These facial expressions are mainly analyzed through promising techniques of computer vision and image processing. As discussed, the emotions are detected by facial expressions interpreted through facial textures, facial muscles, eyes, and eyebrows. The generic model for behavioral analysis is as follows:

1. Finding of user's face from vide frames
2. Extraction of facial features and its normalization to feature vectors
3. Classification of user emotions
4. Calculation of the intensity of each emotion (if required)

In general, it captures the image of patient through some camera. The captured image is processed to detect the face and feature extraction. It then uses the classifier which classifies the image into different emotion classes. This classifier has been pre-trained using a huge training dataset. The approaches for recognition of facial expressions have been broadly classified into geometric based approaches and appearance based approach.

The geometric-based approaches represent the human face is represented in the form of a feature vector that represents the human face in terms of its geometry. This geometry comprises of shapes, points, and locations of facial components like eyes, nose, eyebrows, and mouth and their distances. The downside of this geometric based approach is that it requires an accurate and reliable detection method

(Tivatansakul & Ohkura, 2014). Another challenge for geometric-based approach is that the facial geometry of human is affected by the environment and climatic conditions. For instance, in case of extreme cold weather or some stinking odor, the face geometry changes even in the absence of any emotional transformation.

Appearance-based approaches work on the principle of extracting changes in face appearances and skin textures. There are numerous existing methods to extract changes in face appearance. One such method is Local Binary Patterns (LBP) which divides the image into a grid of rectangular regions. Each region is then encoded into curves, edges and other local features using comparison with neighboring pixels and center value. Thereafter, it constructs a 256 level histogram for each region which are later concatenated to form a global description of entire face (Tivatansakul & Ohkura, 2014). Similarly, there exists a Local Directional Pattern (LDP) method which employs edge detector to evaluate edge response in all eight directions. LDP is a robust method for appearance-based behavioral analysis as LDP uses edge responses which are stable in comparison to intensity based values for generation of binary patterns. Apart from LDP and LBP, there are few other methods in existence which can perform appearance-based behavioral analysis. Researchers have also proposed few Internet of Things (IoT)–based models for automated monitoring and assistance of patients in their homes (Enshaeifar et al., 2018). These kind of systems basically consists of sensors in order to sense vital body parameters. The sensed parameters are then forwarded to some back-end system that implements some analysis method (Bhardwaj, Khanna, Sharma, & Chhabra, 2019). Thereafter, user interface presents the clinical and technical alerts to health profession. With the help of this information, health professionals are capable to monitor the health of patients and accordingly suggest treatment around the clock. Enshaeifar et al. (2018) have employed such approach for maintaining the well-being of dementia affected patients by analyzing their behavior captured through installed camera and other sensing devices. The observed values are processed using data analytics and machine learning algorithms to generate notifications and recommendations to the patients thus providing timely and effective support thus preventing worsening of health.

Further, Ahn, Fox, and Jabon (2010) have proposed a model that analyzes facial expressions beyond categorized measurements. For the same, a computer is installed with a camera, tracking software and ML that enables selection of the most relevant facial features and improves the prediction model in a cost-effective manner. Ability to fit it beyond predefined categories achieves greater power and efficiency. Its working model has advocated its applications in various domains like predicting unsafe behavior of driver, monitoring operator fatigue and shopping experience etc. Similarly, Smirnov, Banger, Davis, Muraleedharan, and Ramachandran (2013) have also proposed a two stages framework for emotion recognition through facial feature extraction. These two steps in this framework are signal acquisition (through camera) and signal processing. In signal processing, the image is transformed to two dimensional array which is used for face detection and feature extraction. The extracted features are further classified into different emotions using various ML algorithms. Sapiński, Kamińska, Pelikant, and Anbarjafari (2019) have also proposed a similar model to recognize human emotion using body movement based on body joints within the tracked Skelton.

13.3 Comparative Analysis of Various Perspectives

In the previous section, we have discussed various perspectives to analyze emotional health. Each of these perspectives has its own capabilities and limitations. Text processing is a better perspective as it requires less storage in comparison to its counterparts. In speech, the voice signals are processed to recognize and extract different emotions. On the contrary, video data processes audio, image, and text (sometimes) thus giving better results than other approaches. Although, video processing gives better results, its processing takes a lot of resources in terms of time and memory.

Apart from requirement of resources, these approaches perform best in some specific application scenarios. For instance, facial expression and speech can be employed in everyday life where user interacts in most natural way as it does not require any sensor (Tivatansakul & Ohkura, 2014). Moreover, facial expression overcome the barrier of cross-culture analysis as different languages and geographic regions have different dialects and voice quality, whereas facial expressions are universal to a certain extent. Facial expression processing also achieves consistent and higher accuracy in comparison to speech processing despite regional and language discrepancies. Moreover, the technological advancement has propelled the availability of image capturing devices in an affordable range, further enhancing the popularity of facial expressions perspective. However, mere facial expression processing is not sufficient in some scenarios. One such example is healthcare where it is mandatory to analyze biological signals in addition to facial expressions (Tivatansakul & Ohkura, 2014). The behavioral method obtains enhanced accuracy as mentioned earlier. However, it also has some challenges. The most critical challenge is its high computational complexity due to which it requires high computational time (Thacker & Makwana, 2019), requires high training time and thus not suitable for very large datasets. Another concerning issue is that behavioral method experiences difference based on gender, culture, geographical region and environmental conditions.

Similar to facial expressions processing, text processing also has some associated challenges. The foremost challenge of text processing is extraction of opinion from text understanding the context of words. Another major challenge for text processing is subjectivity detection as some text may be sensitive for some while others may feel quite neutral toward the same. Subjectivity detection for small message like Twitter data is further cumbersome due to lack of contextual information and requirement of suitable regularization to fill missing data. Context dependency is another major challenge for text processing as a particular word could have different subjectivity in a particular context (Chaturvedi, Cambria, Welsch, & Herrera, 2018). For instance, the adjective "long" have positive reflection in "long battery life" and negative in "long waiting time."

Also, emotional health analysis through speech processing has some associated challenges. Noise robustness is a serious challenge specially in changing acoustic environments. Thus, recording of natural emotion is a challenging task as even the most popular recording protocols lacks recording elicited emotion (Ratna Kanth & Saraswathi, 2014). Another challenge is the possibility of multiple emotions in same utterance. The efficiency of speech processing is also affected as expression of a certain emotion by a person is also influenced by his or her background (culture, region, etc.). Resultantly, same emotion may be expressed differently by different speakers. Sometimes, a speaker a may undergo a particular emotional state for a prolonged period (days, weeks, or months). In such scenario, other emotions will not last long and thus it is not evident which emotion will be recognized by the recognizer among long-term or transient emotion.

13.4 Case Studies

Esturgó-Deu and Sala-Roca (2010) has implemented emotional analysis in order to analyze the behavior of students in primary education. On the same line, Wong, Wong, and Peng (2010) believed that the Emotional Intelligence (EI) have an impacts on the job outcome for each employee. Thus, authors Wong et al. (2010) in have implemented the approach to analyze effect of teacher's EI on her job satisfaction. The analysis of emotional health is also implemented for evaluating emotional health of employees in workplace. The study of emotional health of employees is based on the principle that that organizations involve complex and competitive relationships. It is also required as the employees generally need to interact with some seniors and peers not of their choice which has a significant impact on their emotional health. The evaluation of employees at workplace is required as it helps to maintain his emotional status at optimum levels thus creating a healthy workplace environment for all others. On the other hand, if emotional health of an employee depletes, it results in his tendency to physically and psychologically withdraw from work by engaging him in some non-work related activities. This also results in a decrease in his performance as the employee experiences a downfall in self-efficacy and escalation in stress and tension.

Extending the research further, Sullivan (2018) have analyzed the group emotions in order to research the impact of sporting events. Similarly, Raman, Sambasivan, and Kumar (2016) have attempted to study the impact of emotional health on counterproductive work behavior (CWB) of government employees. It is observed that CWB instantly negatively impacts the customer's association with the organization. Moreover, researchers showed that this CWB of employees may results into huge costs to organizations sometimes running into billions of dollars.

13.5 Conclusion and Future Work

In this chapter, authors have presented the various perspectives of emotional health analysis. Authors have also discussed the urgency to evaluate emotional health of humans by discussing the case studies related to students, employees, government officials, etc. Each perspective for emotional health analysis has its own constraints and challenges. For instance, performance of speech processing is greatly influenced by the dialect, region, and gender of the subject. Similarly, behavioral analysis is also influenced by gender and climatic conditions thus impacting the accuracy of model. Thus, each model has its own limitations.

In order to enhance the efficiency and accuracy of such system, it must take the research one step ahead by integrating multiple approaches. The research must also be extended in the direction of incorporating nature of stimulus for change in emotional status (Schirmer & Adolphs, 2017). For instance, the research must be carried out in the direction of dynamic stimulus like touch and vocalization that resulted in change in emotional state. It may also be taken further to study the location of touch on body or its cultural significance. Thus, it is concluded that visual, auditory and tactile senses offer specific strengths and challenges for emotional analysis. The research can be carried forward in this direction to improve the accuracy and efficiency of existing approaches for holistic and ecological emotional health analysis.

References

Ahn, S. J., Bailenson, J., Fox, J., & Jabon, M. (2010). 20 Using automated facial expression analysis for emotion and behavior prediction. *The Routledge Handbook of Emotions and Mass Media*, 349.

Ambarkar, S., & Akhare, R. (2020, February). A study to analyze impact of social media on society: WhatsApp in particular. *International Journal of Education and Management Engineering, 10*(1), 1–10. https://doi.org/10.5815/ijeme.2020.01.01

Bhardwaj, K. K., Khanna, A., Sharma, D. K., & Chhabra, A. (2019). *Energy conservation for IoT devices* (Vol. 206). Singapore: Springer. https://doi.org/10.1007/978-981-13-7399-2

Bobicev, V., Sokolova, M., Jafer, Y., & Schramm, D. (2012). Learning sentiments from tweets with personal health information. *Lecture Notes in Computer Science (Including Subseries Lecture Notes in Artificial Intelligence and Lecture Notes in Bioinformatics), 7310*, 37–48. https://doi.org/10.1007/978-3-642-30353-1_4

Carrillo-de-Albornoz, J., Vidal, J. R., & Plaza, L. (2018). Feature engineering for sentiment analysis in e-health forums. *PLoS One, 13*(11), 1–25. https://doi.org/10.1371/journal.pone.0207996

Chaturvedi, I., Cambria, E., Welsch, R. E., & Herrera, F. (2018). Distinguishing between facts and opinions for sentiment analysis: Survey and challenges. *Information Fusion, 44*(December 2017), 65–77. https://doi.org/10.1016/j.inffus.2017.12.006

Chen, M., Yang, J., Hao, Y., Mao, S., & Hwang, K. (2017). A 5G cognitive system for healthcare. *Big Data and Cognitive Computing, 1*(1), 2. https://doi.org/10.3390/bdcc1010002

Davletcharova, A., Sugathan, S., Abraham, B., & James, A. P. (2015). Detection and analysis of emotion from speech signals. *Procedia Computer Science, 58*, 91–96. https://doi.org/10.1016/j.procs.2015.08.032

Deng, Y., Stoehr, M., & Denecke, K. (2014). Retrieving attitudes: Sentiment analysis from clinical narratives. *CEUR Workshop Proceedings, 1276*, 12–15.

Enshaeifar, S., Zoha, A., Markides, A., Skillman, S., Acton, S. T., Elsaleh, T., … Barnaghi, P. (2018). Health management and pattern analysis of daily living activities of people with dementia using in-home sensors and machine learning techniques. *PLoS One, 13*(5), 1–20. https://doi.org/10.1371/journal.pone.0195605

Esturgó-Deu, M. E., & Sala-Roca, J. (2010). Disruptive behaviour of students in primary education and emotional intelligence. *Teaching and Teacher Education, 26*(4), 830–837. https://doi.org/10.1016/j.tate.2009.10.020

Goeuriot, L., Na, J. C., Kyaing, W. Y. M., Khoo, C., Chang, Y. K., Theng, Y. L., & Kim, J. J. (2012). Sentiment lexicons for health-related opinion mining. In *IHI'12—Proceedings of the 2nd ACM SIGHIT International Health Informatics Symposium* (pp. 219–225). https://doi.org/10.1145/2110363.2110390

Hakak, N. M., Mohd, M., Kirmani, M., & Mohd, M. (2017, July). Emotion analysis: A survey. In *2017 International Conference on Computer, Communications and Electronics, COMPTELIX 2017* (pp. 397–402). https://doi.org/10.1109/COMPTELIX.2017.8004002

Ji, X., Chun, S. A., Wei, Z., & Geller, J. (2015). Twitter sentiment classification for measuring public health concerns. *Social Network Analysis and Mining, 5*(1), 1–25. https://doi.org/10.1007/s13278-015-0253-5

Kerkeni, L., Serrestou, Y., Mbarki, M., Raoof, K., Ali Mahjoub, M., & Cleder, C. (2019). Automatic speech emotion recognition using machine learning. In *Social Media and Machine Learning [Working Title]* (pp. 1–16). Rijeka: IntechOpen. https://doi.org/10.5772/intechopen.84856

Khan, K., & Ejaz, M. (2016, May). *Emotion detection through text IoT big data analysis view project*. Retrieved from www.theinternationaljournal.org

Lim, S., Tucker, C. S., & Kumara, S. (2017). An unsupervised machine learning model for discovering latent infectious diseases using social media data. *Journal of Biomedical Informatics, 66*, 82–94. https://doi.org/10.1016/j.jbi.2016.12.007

Mirheidari, B., Blackburn, D., & Walker, T. (2019). Dementia detection using automatic analysis of conversations. *Computer Speech and Language, 53*, 65–79.

Nwe, T. L., Foo, S. W., & De Silva, L. C. (2003). Speech emotion recognition using hidden Markov models. *Speech Communication, 41*(4), 603–623. https://doi.org/10.1016/S0167-6393(03)00099-2

Ramalingam, V. V., Pandian, A., Jaiswal, A., & Bhatia, N. (2018). Emotion detection from text. *Journal of Physics: Conference Series, 1000*(1), 012027. https://doi.org/10.1088/1742-6596/1000/1/012027

Raman, P., Sambasivan, M., & Kumar, N. (2016). Counterproductive work behavior among frontline government employees: Role of personality, emotional intelligence, affectivity, emotional labor, and emotional exhaustion. *Revista de Psicologia Del Trabajo y de Las Organizaciones, 32*(1), 25–37. https://doi.org/10.1016/j.rpto.2015.11.002

Ratna Kanth, N., & Saraswathi, S. (2014). A survey on speech emotion recognition. *Advances in Computer Science and Information Technology (ACSIT), 1*(3), 135–139. Retrieved from http://www.krishisanskriti.org/acsit.html

Mohanty, S. N., Pratihar, D. K., & Suar, D. (2015). Study on influence of mood states on information processing during decision making using fuzzy reasoning tool and neuro-fuzzy system developed based on Mamdani approach. *International Journal of Fuzzy Computing Modeling, 1*(3), 252–268.

Mohanty, S. N., & Suar, D. (2013). Decision-making in Positive and negative prospects: Influence of certainty and affectivity. *International Journal of Advances in Psychology, 2*(1), 19–28.

Mohanty, S. N., & Suar, D. (2014). Decision Making under uncertainty and information processing in positive and negative mood states. *Psychological Reports, 115*(1), 91–105.

Sambana, B. (2017). Internet of things: Applications and future trends. *International Journal of Innovative Research in Computer and Communication Engineering, 5*(3), 5194–5202. https://doi.org/10.15680/IJIRCCE.2017

Sapiński, T., Kamińska, D., Pelikant, A., & Anbarjafari, G. (2019). Emotion recognition from skeletal movements. *Entropy, 21*(7), 1–16. https://doi.org/10.3390/e21070646

Schirmer, A., & Adolphs, R. (2017). Emotion perception from face, voice, and touch: Comparisons and convergence. *Trends in Cognitive Sciences, 21*(3), 216–228. https://doi.org/10.1016/j.tics.2017.01.001

Shivhare, S. N., & Saritha, S. K. (2014). Emotion detection from text documents. *International Journal of Data Mining & Knowledge Management Process, 4*(6), 51–57.

Smirnov, D., Banger, S., Davis, S., Muraleedharan, R., & Ramachandran, R. P. (2013). Automated human behavioral analysis framework using facial feature extraction and machine learning. *Conference Record—Asilomar Conference on Signals, Systems and Computers, 4*, 911–914. https://doi.org/10.1109/ACSSC.2013.6810420

Sullivan, G. B. (2018, August). Collective emotions: A case study of South African pride, euphoria and unity in the context of the 2010 FIFA World Cup. *Frontiers in Psychology, 9*, 1–18. https://doi.org/10.3389/fpsyg.2018.01252

Thacker, C. B., & Makwana, R. M. (2019). Human behavior analysis through facial expression recognition in images using deep learning. *International Journal of Innovative Technology and Exploring Engineering, 9*(2), 391–397. https://doi.org/10.35940/ijitee.b6379.129219

Tivatansakul, S., & Ohkura, M. (2014). *Emotional healthcare system* (pp. 41–46). New York: IEEE.

Vij, A., & Pruthi, J. (2018). An automated psychometric analyzer based on sentiment analysis and emotion recognition for healthcare. *Procedia Computer Science, 132*, 1184–1191. https://doi.org/10.1016/j.procs.2018.05.033

Wong, C. S., Wong, P. M., & Peng, K. Z. (2010). Effect of middle-level leader and teacher emotional intelligence on school teachers' job satisfaction. The case of Hong Kong *Educational Management Administration and Leadership, 38*(1), 59–70. https://doi.org/10.1177/1741143209351831

Correction to: Theory and Implications of Information Processing

G. Sucharitha, Anjanna Matta, Kanagala Dwarakamai, and Bodepu Tannmayee

Correction to:
Chapter 4 in: S. N. Mohanty (ed.), *Emotion and Information Processing*, https://doi.org/10.1007/978-3-030-48849-9_4

This book was inadvertently published without updating the following error.

Chapter author G. Sucharitha's name was erroneously published in the Front Matter and in Chapter 4.

The above correction has now been updated.

The updated online version of this chapter can be found at
https://doi.org/10.1007/978-3-030-48849-9_4

S. N. Mohanty (ed.), *Emotion and Information Processing*,
https://doi.org/10.1007/978-3-030-48849-9_14

Index

S. N. Mohanty (ed.), *Emotion and Information Processing*,
https://doi.org/10.1007/978-3-030-48849-9

GPSR Compliance

The European Union's (EU) General Product Safety Regulation (GPSR) is a set of rules that requires consumer products to be safe and our obligations to ensure this.

If you have any concerns about our products, you can contact us on ProductSafety@springernature.com

In case Publisher is established outside the EU, the EU authorized representative is:

Springer Nature Customer Service Center GmbH
Europaplatz 3
69115 Heidelberg, Germany

Batch number: 10844692

Printed by Printforce, the Netherlands